AF649333

LIFE SCIENCES AND SPACE RESEARCH XII

COSPAR

LIFE SCIENCES AND SPACE RESEARCH XII

Proceedings of the Open Meeting of the Working Group on Space Biology of the Sixteenth Plenary Meeting of COSPAR

Constance, F.R.G., 23 May — 5 June 1973

Organized by

THE COMMITTEE ON SPACE RESEARCH — COSPAR

and

THE "DEUTSCHE FORSCHUNGSGEMEINSCHAFT"
OF THE FEDERAL REPUBLIC OF GERMANY

Edited by

P. H. A. SNEATH

AKADEMIE-VERLAG · BERLIN
1974

Executive Editor:
Dr. A. C. Stickland

Library of Congress Catalog Card Number 63—6132

© Akademie-Verlag, Berlin, 1974
All Rights Reserved. No part of this publication may be reproduced, stored in a retrieval system, or transmitted, in any form or by any means, electronic, mechanical, photo copying, recording or otherwise, without the prior permission of the Copyright owner.
202 · 100/473/74
Gesamtherstellung: VEB Druckhaus „Maxim Gorki", 74 Altenburg
Bestellnummer: 761 941 7 (3060/XII) · LSV 1305, 1495
Printed in GDR

Professor W. V. Vishniac
(1922—1973)

It is with deep regret that we record the death of Professor Wolf V. Vishniac by a tragic fall in Antarctica on 10 December 1973 at the early age of 51. Professor Vishniac was conducting investigations on the ability of micro-organisms to survive and multiply under harsh conditions in the Asgard Mountains when he fell down a steep slope.

He had been closely associated with the work of COSPAR for many years: at the time of his death he was a member of the Executive Council of COSPAR (as a representative of the International Union of Biochemistry), co-chairman of the COSPAR Working Group 5 on Space Biology, and of two of its panels, Exobiology and Planetary Quarantine; he was also a member of the Committee of Working Group 7 on Space Research Studies of the Moon and Planets. From 1968 to 1972 he was Scientific Editor for the COSPAR Life Sciences and Space Research volumes.

Professor Vishniac's interests in space research were directed toward the search for extra-terrestrial life, and he was the author of numerous publications on this, many of them in the Life Siences series of COSPAR. He was well known for his design for an ingenious instrument to suck planetary dust into a culture chamber in which micro-organisms could then multiply: changes in turbidity and acidity would then be detected. As might be expected this device was affectionately nicknamed the 'Wolf Trap'. For his work with NASA he received the Apollo Achievement Award, the Lunar Science Award and the Lunar Quarantine Operations Award.

His appreciation of the risks of contaminating the planets with terrestrial micro-organisms led to his staunch support of the planetary quarantine requirements laid down by COSPAR. In this his service was invaluable, both through his demonstrations that micro-organisms could multiply in the most barren natural habitats, such as the dry valleys of Antarctica, and also because of his friendly and persuasive advocacy of international collaboration in investigating such problems and in guarding against the dangers of contamination. His wide knowledge of biochemistry and ecology as well as of microbiology gave him exceptional insight into problems of this kind. COSPAR has many reasons to be grateful for his devoted work on the Working Groups and Panels and his scientific editing.

Professor Vishniac was born in Berlin of Latvian parents, emigrated to the United States of America, and became a naturalized citizen of that country in 1946. He was a member of the Yale University staff from 1952 to 1961 and of the University of Rochester from 1961 to the time of his death. He had been a consultant to NASA since 1965, serving on the Lunar and Planetary Missions Board; for the Office of Manned Space Flight on the Joint USSR/USA Editorial Board on Space Biology and Medicine; and other committees, as well as various lunar research teams, being leader of the Biology Instrument Team. He was also an active member of the Space Sciences Board of the U.S. National Academy of Sciences, and in this connection was linked directly with COSPAR.

Wolf Vishniac was a live and inspiring teacher and colleague, filled with enthusiasm to learn more about the secrets of possible life in space. His knowledge of biology was exceptionally broad, his appreciation of the interactions between different kinds of organism, and between organisms and environment, was quite out of the ordinary for one who had specialized in microbiology and biochemistry. This was due to his wide interests, which were not confined to science, but ranged over art, music and philosophy. His wide knowledge and love of literature, and his great sense of humour, were apparent even in his casual correspondence and conversation. He was an inspiring colleague and a cheerful companion who will be greatly missed.

C. de Jager
President of COSPAR

Preface

The proceedings of the Life Sciences sessions of the 16th Plenary Meeting of the Committee on Space Research (COSPAR), held at Konstanz in May and June 1973, are presented in this twelfth volume of *Life Sciences and Space Research.* They embrace a wide range of topics of both theoretical and practical importance.

There is a continuing trend toward more detailed work upon the biological effects of radiation produced by particles of very high energy, and we are now beginning to learn something about the kind of cellular damage that these particles cause. This work is made possible by advances in technique for recording particle tracks in relation to biological objects, and several studies on this are included in the present volume.

Another topic that is commanding increased attention is the effect of weightlessness upon developmental processes in biology, and the interaction between weightlessness and other effects of space flight. Experiments in this field are apt to be difficult to carry out and it is therefore with pleasure that the scientific community watches the steady progress of our knowledge. Attention continues also upon the important field of planetary quarantine, with the growing realization that all the planets should be protected from unwise contamination.

I would like to express my grateful thanks to Dr. A. C. Stickland for her great help in editing and preparing the volume for press.

It is with great regret that we announce the death of Professor Wolf Vishniac, a former editor of this Series; an appreciation will be found in the volume.

P. H. A. SNEATH

Contents List

Gravitational Biology

Planetary Quarantine

Exobiology

Life Sciences and Space Research XII — Akademie-Verlag, Berlin 1974

DETECTION OF EXTRATERRESTRIAL LIFE BY RADIOMETRIC TECHNIQUES

A. A. IMSHENETSKY and B. G. MURZAKOV

Institute of Microbiology, USSR Academy of Sciences, Moscow, USSR

The evolution of radioactive CO_2 from ^{14}C labelled substrates by desert soils has been studied. Formate, acetate, lactate, glycine and protein hydrolysate are attached much more rapidly than glucose in the first few hours of incubation. Glucose utilization increases considerably after 12 hours incubation.

The rate of $^{14}CO_2$ evolution is much reduced by low humidity. The optimal temperature is 28–37°, and addition of yeast autolysate and liver extract increases $^{14}CO_2$ evolution. Attack on radioactive n^- paraffin was also demonstrated.

The detection of extraterrestrial life is one of the most exciting scientific problems, of both biological and philosophical significance, but recent studies of the planets nearest to the earth still cannot give an unambiguous answer.

Of all the planets of the solar system Mars remains the most probable habitat of living beings. The factor which limits most of all the possibility of life on Mars is undoubtedly a very low content of water in the soil and in the atmosphere. However, we must take into account the possibility of large amounts of water stored in the form of ice in the polar regions. This water could become periodically available for hypothetical inhabitants of Mars.

If we accept the analogy with terrestrial life, life on Mars may be represented by organisms having various adaptive mechanisms enabling them to exist under xerophytic conditions. The detection of such organisms is a complicated task. The main requirement is for techniques that register the growth and vital activity of the cells over time; the absence of these changes in a sterile control, or the cessation of the changes upon addition of an antimetabolite, suggests that the processes are biological ones and therefore the presence of micro-organisms.

The radioisotope techniques can be used in exobiology in several ways and in particular for investigating soil "respiration". Evolution of carbon dioxide is the most important indication of the biological origin of soils. Soil humidity is one of the main factors involved, increased humidity giving increased activity of the soil microflora. In the atmosphere of Mars water is present only in traces. Therefore an attempt was made to detect "soil respiration" at the lowest level of soil humidity necessary for the vital activity of micro-organisms. The limiting factor is the time to obtain a reliable signal, for this is extremely important for the remote control of the experiments.

No soil "respiration" was registered with any concentration of uniformly labelled glucose if the humidity of soil was 4% (Fig. 1). An increase of soil humidity caused an increase in soil "respiration"; the optimum humidity was found to vary with the concentration of radioactive glucose: at a concentration of glucose of 10 μC/g soil the optimal humidity is 18—37%, at 25 μC/g, 12—30%, at 50 μC/g,

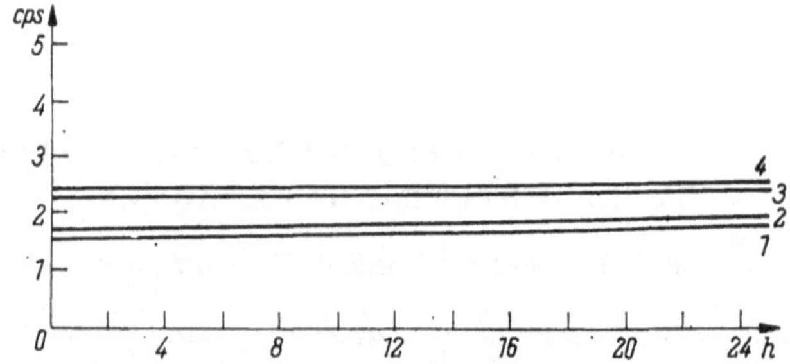

Fig. 1. Evolution of $^{14}CO_2$ as a function of the content of uniformly labelled glucose (1 and 2, 25 μC per g soil; 3 and 4, 100 μC per g soil) and of the temperature of incubation of soil at 4% humidity (1 and 3, 10°; 2 and 4, 37°).

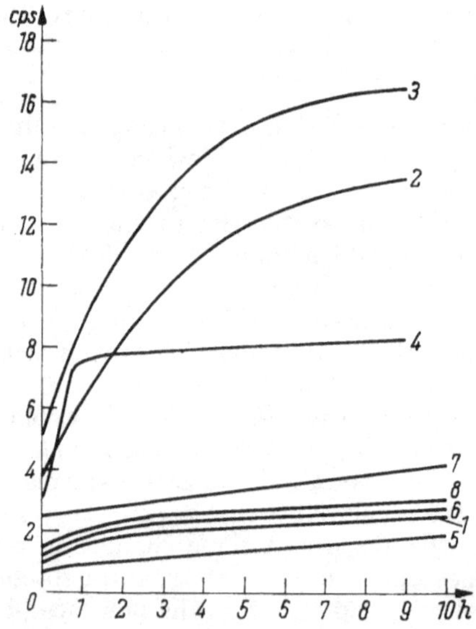

Fig. 2. Evolution of $^{14}CO_2$ as a function of the temperature of incubation and concentration of uniformly labelled glucose in soil with 20% humidity: 1 to 4, 100 μC per g soil; 5 to 8, 25 μC per g soil. Temperatures were: 1 and 5, 10°; 2 and 6, 28°; 3 and 7, 37°; 4 and 8, 50°.

9—31% and at 100 μC/g, 8.5—31%. The radioactivity of the glucose was 6 μC/mg.

Production of carbon dioxide correlates not only with humidity of the soil but also with temperature. Fig. 2 shows that the optimal temperatures are 28° and 37°. An increase of temperature at low humidity does not cause an increase in soil "respiration".

The method can detect 1000 cells of the non-sporeforming micro-organism *Pseudomonas fluorescens* in the soil at humidity of 20 and 36%. A small increase in the signal was found with growing of 1000 cells of the sporeforming micro-organism *Bacillus subtilis* at humidity of 20%. A decrease of soil humidity to 5% shifts the minimum numbers of micro-organisms that can be detected to a value higher than 10^5 cells. Sporeforming bacteria do not show any conspicuous

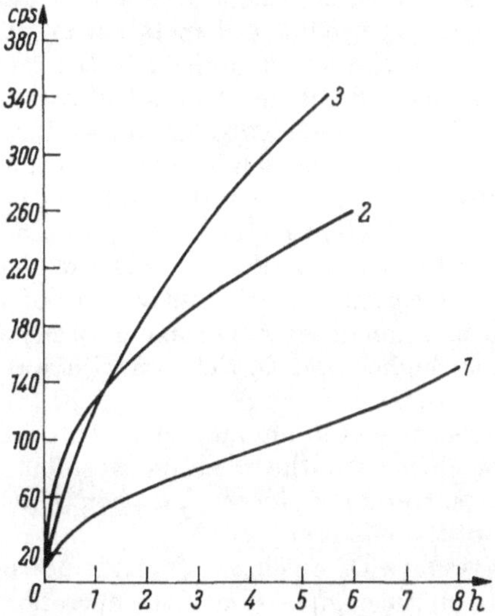

Fig. 3. Rate of soil respiration during incubation of 2 g soil in the presence of 1.7 μC uniformly labelled ^{14}C glucose plus 3.4 μC ^{14}C formate and 4.7 μC ^{14}C glycine. Humidity: 1, 4%; 2, 10%; 3, 20%. The specific radioactivity of the formate and glycine was 72 and 100 microcuries/mg respectively.

resistance to a decrease of soil humidity compared with non-sporeforming micro-organisms.

Glucose is only slowly decarboxylated during the first hours of incubation (see below), so that substrates that are decomposed by the soil microflora were used more quickly in later experiments. Fig. 3 shows the $^{14}CO_2$ evolution during the incubation of 2 g of soil from the Kara-Kum desert (Turkmen SSR) at humidities of 4%, 10% and 20% and containing uniformly labelled glucose plus formate-^{14}C and glycine-1-^{14}C as nutrient substrates. The rate of evolution of labelled carbon dioxide is much higher than when only glucose is used. We have established the important fact that evolution of $^{14}CO_2$ can occur at a humidity of 4%, which is only a few tenths of one per cent higher than the maximal normal soil humidity in the desert. The minimal number of micro-organisms that can be detected at 10% humidity is about 1000 cells/g soil. The minimal number is 5×10^3/g if the humidity is reduced to 4%. The aim of the workers is to maintain, as far as possible, the same conditions as occur on Mars. Cultivation of micro-organisms in a liquid medium might well kill any xerophytes that may inhabit Mars. However, several arguments can be adduced against this argument. First, the evidence of terrestrial ecology shows that organisms that are capable of existing under extreme conditions do nevertheless develop, as a rule, better under less extreme conditions. For instance, psychrophilic bacteria, although they can grow at —6°, grow much better at +18°. There are no xerophytic micro-organisms known which cannot grow in liquid media. Second, moistening of soil in order to get a more intensive signal also produces a considerable deviation from the natural conditions of the environment in the same way that cultivation in a liquid nutrient solution does. However, there are more favourable conditions for microbial growth in the last case (constant flow of fresh nutrient substances, increased diffusion of gaseous products of metabolism, etc.) which is especially important since the content of micro-organisms in the ground of Mars may well be very low. A considerable advantage of liquid growth media is that one can combine several sensors in the same incubation cell so that microbial growth can also be detected by techniques like nephelometry, potentiometry and manometry. Therefore, the first automatic biological stations should be properly equipped so as to carry out inoculation of soil into a medium containing labelled substrates.

The previous experiments using uniformly labelled glucose were suitable for working out the optimal conditions for cultivation of micro-organisms from desert soils, and for the solution of several problems of metabolism. If the number of counts was 2—3 times higher than in the control it was regarded as a reliable indication of micro-organisms.

A direct correlation between the amount of carbon dioxide evolved and the content of radioactive glucose in the medium was found only within certain limits. The optimal concentration of glucose was about 8 μC/ml which corresponds to 0.85 μM glucose/ml of the medium.

The shape of the curve of CO_2 evolution depends not only on the number of inoculated cells but also on the composition of the microflora. A reliable indication of the presence of active microflora can be obtained after 1.5—2 hours when

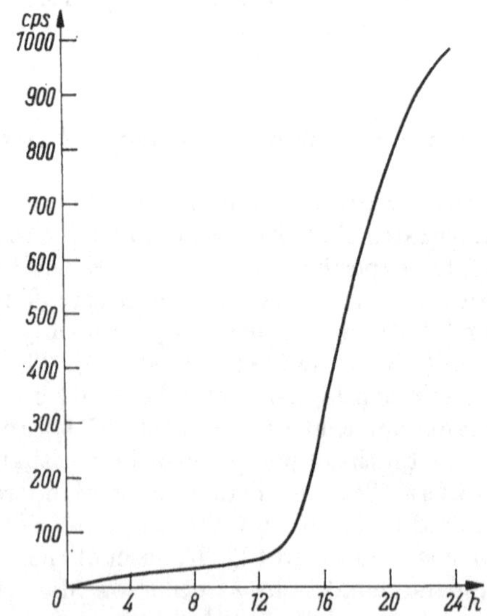

Fig. 4. Rate of $^{14}CO_2$ evolution during incubation of 500 mg of desert soil in medium containing 8.3 μC/ml uniformly labelled glucose.

glucose is used as a carbon source but a considerable evolution of carbon dioxide is detected only after prolonged incubation (Fig. 4). This may be due to complicated ecological and physiological interrelations within a microbiocenoses of desert soils. It is possible that glucose is rapidly utilized by the micro-organisms as a source for the construction of their protoplasm especially during the first hours of incubation. It is also possible that microbial growth under the severe conditions found in desert soils which contain little nutrients is such that the micro-organisms more readily consume simpler organic compounds than glucose. Therefore a major

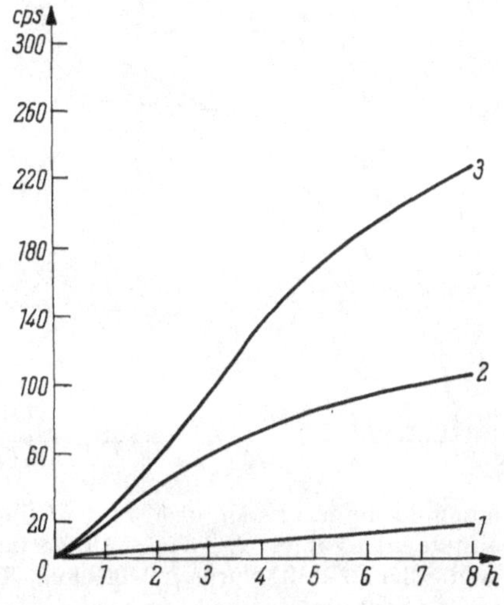

Fig. 5. Rate of $^{14}CO_2$ evolution during incubation of 500 mg of desert soil in medium containing: 1, 8.3 μC 1-6-^{14}C-glucose/ml medium; 2, 4.15 μC sodium lactate-1-^{14}C/ml medium; 3, 8.3 μC sodium lactate-1-^{14}C/ml medium. The specific radioactivity of the glucose and lactate was 6 and 64 μC/mg respectively.

condition for obtaining reliable data on the metabolic activity of microflora is to use a medium that is capable of meeting the physiological requirements of as wide a range of soil micro-organisms as possible. This approach was used by Oyama [1] who suggested a complex medium containing various amino acids, vitamins, cofactors, organic acids and mineral salts. The majority of these compounds are contained in yeast autolysate and liver extract which have been suggested as additives to the medium N 11 [2] and which was used for studying the dynamics of decomposition of 1-6-^{14}C-glucose [3].

Evolution of carbon dioxide from cultures was found to be faster when organic salts, and not glucose, were used as a source of labelled carbon. Fig. 5 shows the curve of $^{14}CO_2$ evolution from labelled sodium lactate. The rate of decomposition of this substrate by micro-organisms from desert soils is much higher than for glucose. The rate of decomposition of sodium formate and acetate is even higher (Figs. 6 and 7). With both glucose and acetate the curve is lower than that for acetate alone. Glucose, although stimulating the production of biomass, seems to inhibit acetate decarboxylation. A similar relationship was shown by incubation of desert soils in medium containing ^{14}C formate (Fig. 7). In the presence of glucose formate $^{14}CO_2$ evolution decreases. Addition of yeast autolysate and liver extract to mineral medium was found to stimulate formate decarboxylation at the beginning of incubation and to hasten the onset of the stationary phase.

In all the above experiments asparagine was used as a source of organic nitrogen.

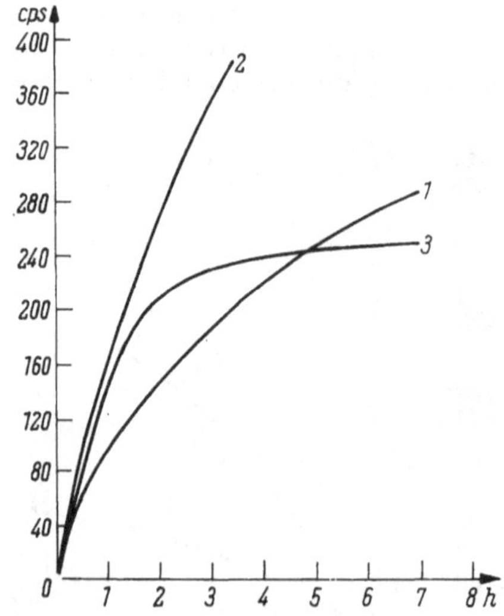

Fig. 6. Rate of $^{14}CO_2$ evolution during incubation of 500 mg of desert soil in medium containing: 1, 4.15 μC 1-^{14}C-acetate/ml medium; 2, 8.3 μC 1-^{14}C-acetate/ml medium; 3, 8.3 μC 1-^{14}C-acetate/ml medium and 8.3 μC 1-6-^{14}C-glucose/ml medium. The specific radioactivity of the acetate was 51 μC/mg.

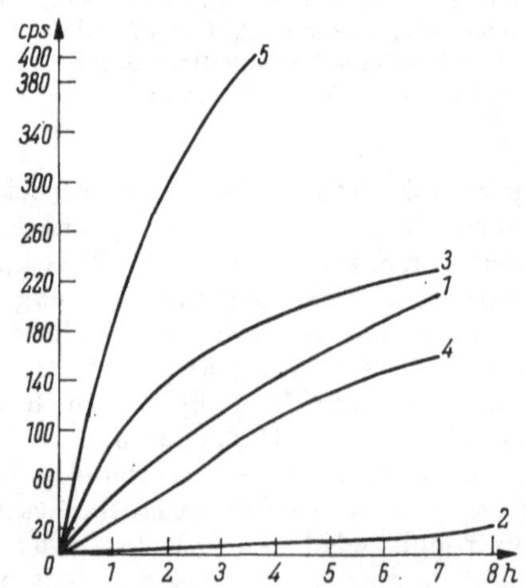

Fig. 7. Rate of $^{14}CO_2$ evolution during incubation of 500 mg of desert soil in medium containing: 1, 8.3 μC ^{14}C-formate/ml medium; 2, 8.3 μC 1-6-^{14}C glucose/ml medium; 3, 8.3 μC ^{14}C-formate/ml medium and yeast autolysate and liver extract; 4, 8.3 μC ^{14}C-formate and 8.3 μC 1-6-^{14}C-glucose in 1 ml of the medium; 5, 4.15 μC ^{14}C-formate and 4.15 μC-1-^{14}C-acetate in 1 ml of the medium.

In one experiment asparagine was substituted by 1-^{14}C glycine which was also the sole source of carbon. Fig. 8 shows the $^{14}CO_2$ evolution; its rate exceeds that found with glucose. However, as expected, the curve is lower than that obtained with uniformly ^{14}C labelled protein hydrolysate. This set of experiments shows that there is a series of organic substrates which are decomposed more rapidly by the microflora of desert soils than by glucose. The best medium which can be

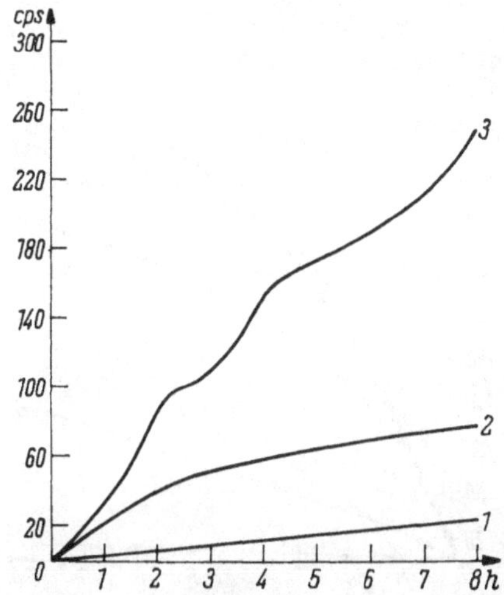

Fig. 8. Rate of $^{14}CO_2$ evolution during incubation of 500 mg of desert soil in medium containing: 1, 38. μC 1-6-^{14}C-glucose/ml medium; 2, 8.3 μC-1-^{14}C-glycine/ml; 3, 8.3 μC/ml of uniformly ^{14}C labelled protein hydrolysate.

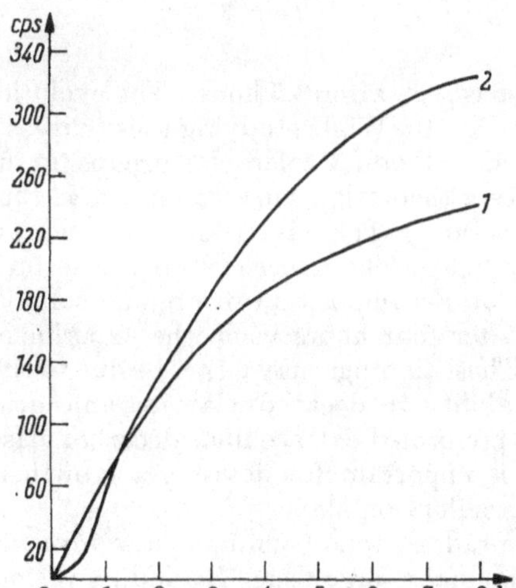

Fig. 9. Rate of $^{14}CO_2$ evolution during incubation of 500 mg of desret soil in medium containing: 1, 1.7 μC 1-6-^{14}C glucose and 3.4 μC formate-^{14}C and 4.7 μC ^{14}C-1-glycine in 1 ml of the basic mineral medium 11; 2, the same substrates plus yeast autolysate and liver extract.

used in exobiological experiments seems to be a combination of several substrates. Fig. 9 presents the curve of $^{14}CO_2$ evolution after inoculation of 500 mg of desert soil into medium containing 1.7 μC of 1-6-^{14}C-glucose, 3.4 μC of formate-^{14}C and 4.7 μC of glycine-1-^{14}C in 1 ml of medium.

The most favourable conditions for decarboxylation are when labelled formate, acetate and glucose are present together in equal amounts. The rate of $^{14}CO_2$ evolution is high from the first few minutes of the experiment. A smooth transition

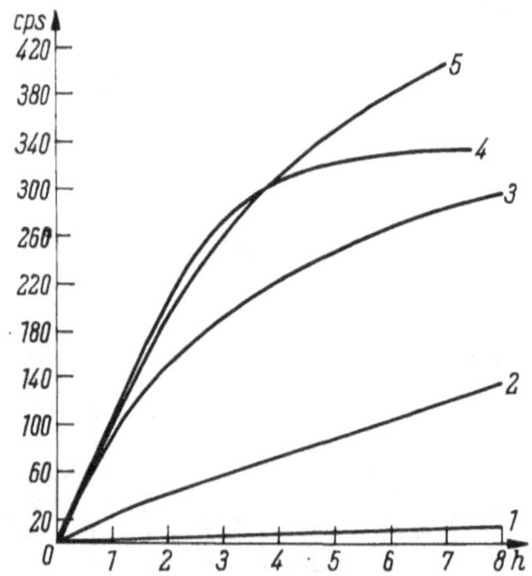

Fig. 10. The rate of $^{14}CO_2$ evolution during incubation of 500 mg of desert soil in medium containing: 1, 4.15 μC uniformly labelled ^{14}C glucose/ml; 2, 4.15 μC ^{14}C formate/ml; 3, 4.15 μC ^{14}C 1-acetate/ml; 4, 4.15 μC/ml of each of the following: uniformly labelled ^{14}C glucose, ^{14}C formate and ^{14}C 1-acetate; 5, the theoretical curve obtained by summation of curves 1, 2 and 3.

to the stationary phase is seen after 3.5 hours. The evolution is higher than with individual substrates (Fig. 10). While studying these curves we become interested in the specialization of individual microbial groups in utilizing various substrates. Fig. 10 includes a theoretical curve obtained as the sum $^{14}CO_2$ evolution from each substrate separately. There is striking agreement with the experimental curve in the first four hours. The experimental curve during this period seems to depend completely on decomposition of organic salts. The effect of glucose becomes pronounced after four hours when the experimental curve falls below the theoretical curve. These findings may either be due to the occurrence of micro-organisms with high ability to decarboxylate organic acids rapidly, or to the presence in the soil of preformed extracellular decarboxylases. Elucidation of the mechanisms involved is important for devising a growth medium for detection of the hypothetical microflora on Mars.

The next step in the radioisotope technique is to investigate the physiological properties of the Martian micro-organisms. By adding various ^{14}C-labelled organic compounds, one can get an idea of the physiological microbial groups inhabiting the planet. In this connection paraffin-decomposing micro-organisms are of interest. They are widely distributed in terrestrial soils. Soil bacteria, fungi and yeasts capable of decomposing aliphatic long chain hydrocarbons are well known.

We have investigated the decomposition of paraffins by the microflora of desert soils. ^{14}C 1-*n*-octadecane (0.83 μC/ml or 0.125 mg/ml) was used as the sole source of carbon in the medium to which was added 500 mg of Kara-Kum soil. The evolution of $^{14}CO_2$ considerably exceeds that from the decomposition of ^{14}C-glucose (Fig. 11). This is unexpected because, although complete decomposition of paraffins is known to take place when oxygen is freely available, in our experiments

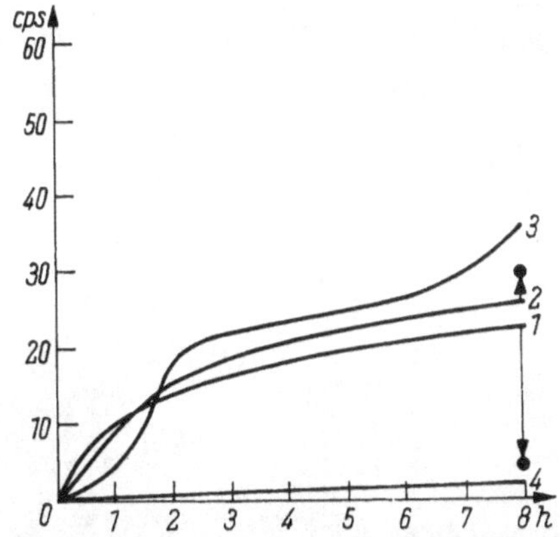

Fig. 11. Evolution of $^{14}CO_2$ on incubation of 500 mg of desert soil in media containing labelled paraffin: 1, 1.0 μC/ml of ^{14}C-1-*n*-octadecane incubated at 28° under restricted aeration; 2, 1.0 μC/ml of ^{14}C-1-*n*-octadecane incubated at 28° under full aeration; 3, 1.0 μC/ml of ^{14}C-1-*n*-octadecane plus 8.3 μC of uniformly labelled glucose, incubated at 28° under restricted aeration; 4, control, with soil previously sterilized, and conditions as for 1.

the aeration was restricted. It is possible that partial breakdown to lower volatile acids may explain our findings.

The radiometric technique can be adapted to testing for decomposition of other substances such as protein and cellulose, and this greatly extends its possibilities for exobiology. Other methods will no doubt be used as well, but the radiometric technique will occupy one of the key positions at all stages of detection and investigation of extraterrestrial forms of life.

References

[1] V. A. Oyama, Icarus **16**, 167 (1972).

[2] L. A. Kuziurina and V. M. Yakshina in: Extraterrestrial Life and Methods for its Detection, Nauka, Moscow 1970 (p. 41).

[3] A. A. Imshenetsky, B. G. Murzakov, A. G. Voskanian and V. K. Surovov, Microbiologia **41**, 727 (1972).

Radiation Biology

Life Sciences and Space Research XII — Akademie-Verlag, Berlin 1974

CHEMICAL PROTECTION AGAINST RADIATION-INDUCED GENETIC DAMAGE DURING THE PERIOD OF AFTER-EFFECTS OF GRAVITY STRESS

M. D. Pomerantseva, V. V. Antipov, G. A. Vilkina and B. S. Gugushvili

USSR Academy of Sciences, Moscow, USSR

The protective effect of cystamine (150 mg kg^{-1}) against genetic damage induced by γ-radiation in germ cells of the CBA line of mice (at doses of 100, 300 and 600 r) was studied. Cystamine reduced the frequency of dominant lethal mutations in sperms, spermatids and spermatocytes. The amount of protection depended on the radiation dose. After cystamine the lethal effect of radiation on the whole organism was reduced 2—3 times. After transverse acceleration (15 g over 10 min) the protective effect of cystamine against dominant lethal mutations induced by radiation (300 r) decreased on the average by 1/3. The action of transverse acceleration alone somewhat reduced the radiosensitivity of germ cells. The administration of the protector and acceleration did not affect the frequency of reciprocal translocations induced by radiation in spermatogonia.

1. Introduction

Space exploration increased the interest in the combined effect of ionizing radiation and different factors of space flight on organisms and for investigating the effectiveness of radioprotectors under these conditions. Several works have been published on the effectiveness of various aminothiols when the action of transverse accelerations and γ-rays are combined [1, 2]. However, in these investigations the protective activity was determined from tests as per cent survival and the life span of irradiated animals. As far as we could find out there is no information on the effect of acceleration on the degree of protective effect of radioprotectors against genetic damage, indeed there is little information on the genetic effects of acceleration and its influence on the mutagenic effect of radiation, and this evidence is often contradictory.

Many chemical protectors, although they protect organisms from the lethal effect of radiation, are significantly less effective or not effective at all in protecting germ cells of mammals. The main reason for the low effectiveness of protectors on germ cells seems to be the fact that they are accumulated in gonads in considerably lower concentration than in other organs. The degree of protection of germ cells of mice from dominant lethal mutations depends both on the radiation dose and on the stage of the germ cell [3, 4]. Protection of premeiotic cells is of special significance, since damage occurring in these cells is retained during the whole reproductive period.

In the present investigation cystamine was chosen as a protector, as it is clinically a highly effective radioprotector and is able to protect germ cells of male mice from radiation-induced dominant lethal mutations [4]. In studying sperms, spermatids and spermatocytes, the frequency of dominant lethal mutations served as the test of protection, and in studying spermatogonia the frequency of reciprocal translocations was used.

2. Materials and Methods

CBA male mice 2.5—3 months old were exposed to total γ-radiation ^{60}Co (100 r, 300 r and 600 r at a dose rate of 30 r/min). Transverse acceleration (15 g over 10 min) was produced in a centrifuge. Cystamine dihydrochloride was administered to animals intraperitoneally in a dose of 150 mg kg^{-1} 10—15 minutes before radiation. The interval between acceleration and radiation was 5—10 min. 1.5—2 hours after radiation each of the males was mated with 3—5 non-irradiated white outbred females for up to one week. With this schedule of mating, the females were fertilized by sperms, which at the time of radiation and acceleration were at the stage of mature sperms (1—7 days), late spermatids (3—14 days), early spermatids (15—21 days), late spermatocytes (22—28 days), early spermatocytes (28—35 days) and spermatogonia (35—42 days). Pregnant female mice were killed and dissected on the 18th—19th day after mating, the number of corpora lutea in their ovaries was counted as well as the number of implantations and the number of living embryos in the horns of the uterus. In order to calculate the frequency of dominant lethal mutations the survival of embryos was determined (the ratio of living embryos to the number of corpora lutea) as well as the pre-implantation losses (the ratio of the difference between the number of corpora lutea and implantations to the number of corpora lutea) and the amount of post-implantation death of embryos (the ratio of the number of dead embryos to the number of implantations).

The amount of induced lethality was calculated according to the following formula:

$$1 - \frac{\%\text{ of survival in the experiment}}{\%\text{ of survival in the control}} \times 100$$

The coefficient of protection was determined according to the following formula:

$$1 - \frac{\%\text{ of induced lethality with a protector}}{\%\text{ of induced lethality without a protector}}$$

The frequency of reciprocal translocations in spermatogonia was determined 3 months after irradiation. The number of reciprocal translocations was analysed in spermatocytes at the stage of diakinese-metaphase of the first meiotic division in air-dried preparations. At least 200 metaphases from each male mouse were analysed.

3. Results and Discussion

The protective effect of cystamine was found only in the group of animals irradiated with 300 r. When cystamine was administered the survival of embryos was statistically significantly higher than controls when cystamine was protecting sperms, spermatids and spermatocytes. Dominant lethals can cause the death of embryos both before and after implantation. The analysis has shown that the level of pre-implantation losses is somewhat lower after irradiation of sperms and late spermatids. The number of pre-implantation losses reflects not only the level of the death of zygotes, caused by dominant lethals, but also the frequency of

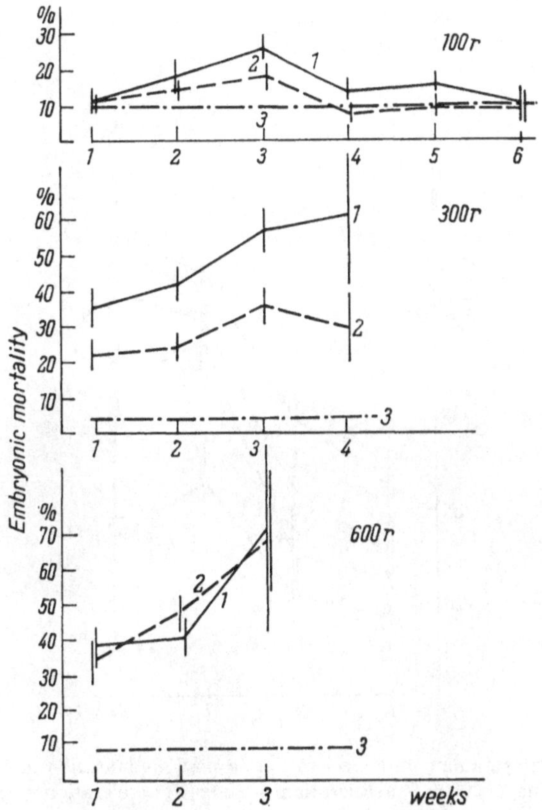

Fig. 1. The effect of cystamine on post-implantation mortality of embryos; 1, radiation; 2, cystamine + radiation; 3, control, not irradiated.

non-fertilized eggs. The increase of survival of embryos, when using the protector, is mainly determined by a decrease in post-implantation mortality of embryos. A distinct protective effect is seen with irradiation of all the stages of spermatogenesis (Fig. 1).

After a dose of 100 r cystamine did not significantly affect the survival of embryos or pre-implantation mortality. Post-implantation mortality of embryos was somewhat lower after cystamine in irradiation of all the stages of spermatogenesis. Taking the stages separately this decrease was not statistically significant but taken together the differences were statistically significant (Fig. 1). After a dose of 600 r cystamine did not affect the frequency of induced dominant lethal mutations (Fig. 1). Thus the degree of protection by cystamine depends on the radiation dose.

It is important to find out whether the protective effect of cystamine is preserved when germ cells are irradiated in the period immediately after acceleration. Fig. 2 shows the results on the influence of acceleration and cystamine on the frequency of dominant lethal mutations in germ cells of mice irradiated with 300 r.

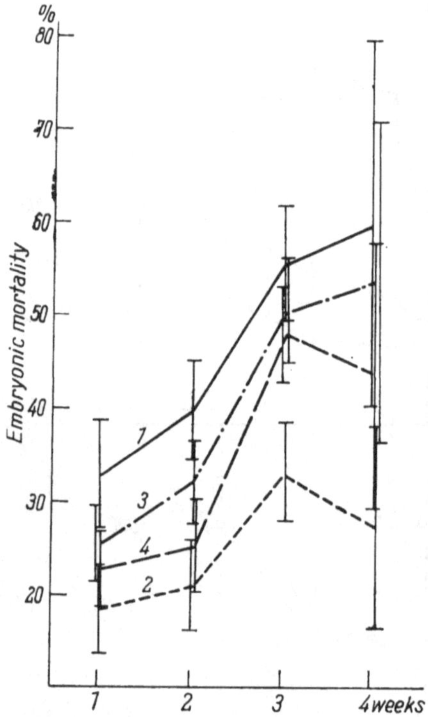

Fig. 2. The effect of cystamine and acceleration on post-implantation mortality of embryos; 1, 300 r; 2, cystamine + 300 r; 3, acceleration + 300 r; 4, acceleration + cystamine + 300 r.

The use of cystamine decreased the number of radiation-induced dominant lethal mutations at all the stages of spermatogenesis. Thus, the protector reduced the lethality for embryos (obtained as a result of crosses between irradiated male mice and non-irradiated female mice), by 17.4 ± 4.1% when affecting mature sperms, by 23.1 ± 3.8% when affecting late spermatids, by 16.9 ± 3.8% when irradiating early spermatids and by 21.8 ± 9.0% when affecting spermatocytes. The average decrease over the whole cycle of spermatogenesis was 17.2 ± 2.3%. The coefficient of protection varied from 0.26 to 0.45 for different stages; the average for all the stages was 0.33. Fig. 2 shows that after acceleration the amount of protection by cystamine was reduced for all stages of spermatogenesis. This reduction was especially pronounced with early spermatids and spermatocytes. For all periods of observation the protective effect after acceleration decreased on the average by about 1/3 (the coefficient of protection reduced from 0.33 to 0.20).

The effect of acceleration also reduced the frequency of radiation-induced dominant lethal mutations (Fig. 2). On the average for the whole cycle the induced lethality decreased by 6.9 ± 2.2% and the coefficient of protection was 0.13. This tendency towards reduction of induced lethality was observed at all stages of spermatogenesis, but statistically significant differences were observed only for the stage of mature sperms. The analysis has shown that on the whole the differences are statistically significant ($P < 0.01$). No statistically significant increase of induced lethality of embryos was observed after acceleration. At the same time post-implantation mortality of embryos was somewhat higher (by 2.6 ± 0.8%) in this group.

Testing for dominant lethal mutations when studying spermatogonia is not convenient, as their frequency is very low and clear dependence of the effect on radiation dose is absent. A better test for the study of radiosensitivity of spermatogonia is the frequency of reciprocal translocations, for which the presence of a linear dependence has been shown. Table 1 shows the data on the influence of

Table 1

The Effect of Cystamine and Acceleration on the Frequency of Radiation-Induced Reciprocal Translocations in Spermatogonia

Treatment	Number of: Males	Number of: Metaphases	Metaphases with no. of translocations: 1	2	3	Total No.	%	Total No. of translocations	Translocations per cell (%)
100 r	16	3160	46	2	—	48	1.51	50	1.58 ± 0.33
Cystamine + 100 r	14	2803	29	1	—	30	1.07	31	1.10 ± 0.24
300 r	10	1982	69	5	—	74	3.73	79	3.98 ± 0.62
Cystamine + 300 r	9	1800	69	2	—	71	3.94	73	4.05 ± 0.94
Acceleration + 300 r	10	2002	72	1	—	73	3.65	74	3.70 ± 0.58
Acceleration + cystamine + 300 r	10	2004	90	4	—	94	4.69	98	4.89 ± 0.66
600 r	17	3205	132	7	1	140	4.36	149	4.65 ± 0.45
Cystamine + 600 r	16	3210	107	16	—	123	3.83	139	4.33 ± 0.89
Control	15	3018	3	—	—	3	0.10	3	0.10 ± 0.05
Cystamine	13	2489	5	—	—	5	0.20	5	0.20 ± 0.09
Acceleration	8	1490	1	—	—	1	0.07	1	0.07 ± 0.02

cystamine and acceleration on the frequency of reciprocal translocations in spermatogonia of mice irradiated with 100 r, 300 r and 600 r. It is seen from the table that at all radiation doses the radioprotector did not significantly affect the frequency of reciprocal translocations. Preliminary accelerations also did not affect the frequency of reciprocal translocations (with a dose of 300 r). This lack of protection by cystamine on spermatogonia can be explained by the selective death of part of these cells because their radiosensitivity is comparatively high (LD ~100 r).

The reasons for the reduction of protection by cystamine after acceleration are not yet clear. It seems to be determined by disturbances of hemodynamics and a slower penetration of protective substances into organs. The protective effect of acceleration against radiation-induced dominant lethals seems to be determined by the fact that centrifugation causes acute hypoxia as a result of redistribution of blood in organs with consequent disturbance of the oxygen supply to the tissues [6]. It shoud be pointed out that the mechanism of the effects of inertial forces is very complicated and uncertain, and perhaps, besides hypoxia, it is connected with the level of accumulation of endogenic sulphydryl groups and changes in the activity of various regulating systems [7].

References

[1] P. P. SAKSONOV, V. V. ANTIPOV and B. I. DAVYDOV, Problemy Kosmicheskoi Biologii **9**, 181 (1968).
[2] V. V. ANTIPOV, M. V. VASIN, B. I. DAVYDOV and V. S. SHASHKOV, Izv. Akad. Nauk USSR 434 (1969).
[3] M. D. POMERANTSEVA, Genetica **3**, 102 (1967).
[4] B. S. GUGUSHVILI, M. D. POMERANTSEVA and G. A. VILKINA, Genetica **8**, 42 (1972).
[5] E. P. EVANS, G. BRECON and C. E. FORD, Cytogenetics, **3**, 289 (1964).
[6] N. A. GAYDAMAKIN, S. G. KULKIN, B. I. DAVYDOV and V. S. SHASHKOV, Problemy Kosmicheskoi Biologii, **14**, 336 (1971).
[7] L. S. SOUTOULOV and P. P. SAKSONOV, Problemy Kosmicheskoi Biologii, **14**, 314 (1971).

Life Science and Space Research XII — Akademie-Verlag, Berlin 1974

INTERACTION BETWEEN RADIATION EFFECTS, GRAVITY AND OTHER ENVIRONMENTAL FACTORS IN *Tribolium confusum*

C. H. Yang and C. A. Tobias

Donner Laboratory/Lawrence Berkeley Laboratory, University of California, Berkeley, Calif., USA

Multicellular organisms possess homeostatic control systems, which, in responding to changes of the external environment, modify the internal milieu of the organism accordingly, in order to make survival and normal physiological processes possible. This group has studied effects on physiological processes of single and combined environmental factors in the flour beetle, *Tribolium confusum*. Studies included low- and high-LET radiation, gravity compensation, near-weightlessness in space flight, ambient temperature, atmospheric composition, and magnetic field effects on the growth and development of *Tribolium*. For somatic effects, there appears to be a "normal physiological range" for each of these environmental variables; moreover, their effects seem independent of each other. When one of the listed environmental factors is near the limits of normal physiological tolerance, however, marked synergism has been observed between the effects of this factor and other environmental stresses.

Fertilized *Tribolium* eggs showed a differential radiosensitivity to external irradiation, and a linear, dose-effect relationship was obtained, as accelerated heavy nitrogen and oxygen ions were used. Synergism was found when *Tribolium* were irradiated with ionizing radiation and subsequently exposed to temperature either below or above their normal physiological range. Gravity compensation or near-weightlessness in space flight have adverse effects on the development of *Tribolium* when an ambient temperature is near the higher or lower limit of tolerance. Similar results were observed when pupae were exposed to a combination of oxygen, temperature and magnetic field stresses.

1. Introduction

The physical, chemical, and biotic environments on earth have been major factors during the many million years of the evolutionary process which have provided selection pressures resulting in organisms particularly well adapted to our present-day environment. We know that many organisms have become extinct, perhaps due to their limited range of adaptability to environmental extremes.

Space flight and conditions on the surface of other planets have greatly extended the range of environmental variables that man and other earth organisms will be subjected to if we are to conquer the planetary system. It has become necessary to make quantitative studies of our responses to extreme environments with regard to physiological and genetic adaptation. Physical factors of the environment are usually fluctuating; on earth this variation is often within a relatively narrow range and for varying periods of time. Living organisms, usually are either able to maintain their internal environment at a steady state through homeostasis or can adapt to the changing conditions by appropriate internal shifts. The control mechanisms involved in homeostasis have gained much interest among biologists since the time of Claude Bernard. The limitations posed by environmental factors on life, however, are still not very well known. We do not know enough about the tolerance of man to changing environmental parameters for extended exposure, especially his ability simultaneously to withstand stresses from more than one environmental parameter. Typical environmental parameters or factors to which living organisms might have been exposed in space flight have been temperature, radiation, magnetic field, gravity, and varying composition of the ambient atmosphere. It is logical to wonder whether "normal" development of organisms can be altered by a change of one or more of these factors. This question takes on more practical significance as man makes his initial attempts to explore the planetary system in prolonged space flights. Even if man does not migrate to the other planets, it is necessary for him to gain a more complete understanding of his own environment and his reactions to it; this knowledge is essential for his survival.

It has been characteristic of many investigations of the homeostatic system to apply stress for a limited period in such a manner that a single environmental variable only is altered while all other variables are kept constant. This has resulted in prolific knowledge of effects from temperature, radiation, and other variables. It is often assumed that these act independently of each other, and very little quantitative knowledge exists on the interaction of the effects of several environmental variables. Selye [1] has introduced the concept of "stress"; all environmental changes contribute to stress, though it is difficult and in some cases impossible to quantify the interactions. In the broad sense of long-term effects, the stress concept does not account for all effects observed in a space environment; it has little room for the explanation of radiation effects; neither can it account for the effects of different gravitational states on morphogenesis.

We have chosen to study the development of the flour beetle, *Tribolium confusum* through all stages of its life cycle. The effects of environmental variables on this organism fall into two more or less distinct classes: subsequent or long-term effects and short-term effects.

Long-term effects often depend on cell proliferation and regeneration; these involve the genetic apparatus and redundancy of genetic information in somatic cells. Some external agents, for example, radiation, act predominantly on the genetic apparatus. Short-term effects depend more on the already expressed genetic information in living cells or, more generally, on gene expression without the necessity for gene duplication. Temperature changes are likely to produce short-term effects.

During the life cycle of *Tribolium* there are two stages during which the organism is particularly sensitive to environmental variables. One of these is the early development of fertilized eggs into larvae and the other the metamorphosis from pupal to adult stages.

2. Response of Organisms to a Single Physical Stress

2.1. Temperature

Tribolium has a complete metamorphosis, and at 30 °C it takes about five days for fertilized eggs to develop into larvae, about three weeks for larvae to develop into pupae, and about five days for pupae to develop into adults, which can live for one to two years. Using one-day old *Tribolium* pupae under aerobic conditions and without application of radiation or magnetic field, we examined the effects of different temperatures on the development of wings and on eclosion. Methods for obtaining the pupae are reported in detail in [2]. This organism showed a

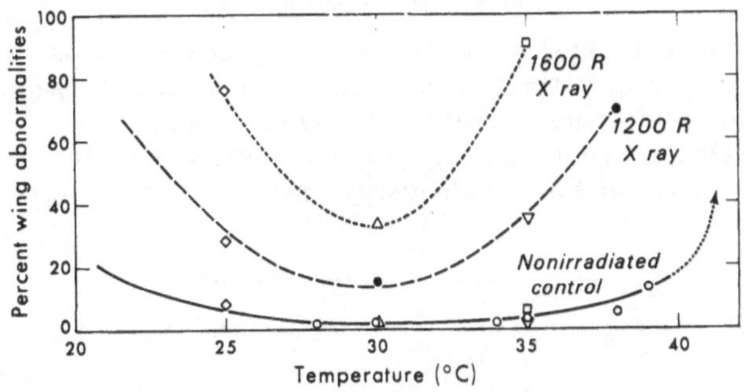

Fig. 1. Effects of different temperatures and radiation on the development of *T. confusum* pupae. 220-kVp X-rays with a dose rate of 1 000 r min^{-1} were used (Phillips 250 kV unit) and the beam was filtered with 1 mm of aluminum.

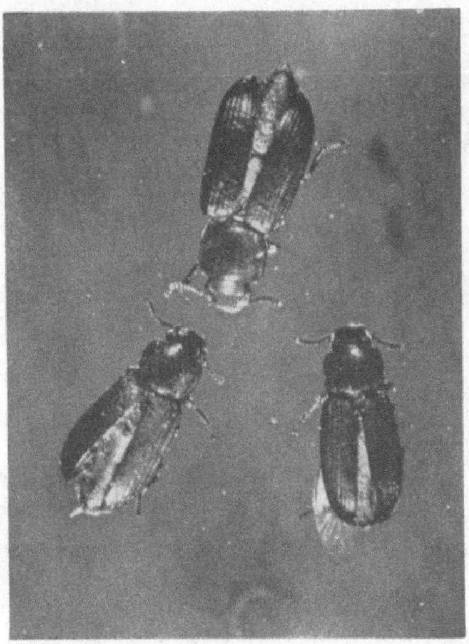

Fig. 2. A picture of wing abnormality of *Tribolium confusum* adult.

minimum failure of wing development within a temperature range of 28 °C to 34 °C (Fig. 1). Lower and higher temperatures seem to be hazardous to this organism as the abnormal development of wings and molting failures occur more frequently. A typical wing deformity is shown in Fig. 2. Results of this experiment suggest that there is a relatively narrow range of temperature within which biological functions can perform with a high efficiency. Outside this optimal temperature range, the efficiency of the biological system decreases and eventually becomes irreversibly damaged.

2.2. Ionizing Radiation

Another important physical factor constantly present in our environment is ionizing radiation from natural sources. Since the earth's atmosphere shields life from most dangerous space radiations, living organisms may lack the ability to cope with the challenge of cosmic rays while in outer space. Our understanding of the biological effects of these high-energy heavy ions is far from complete and

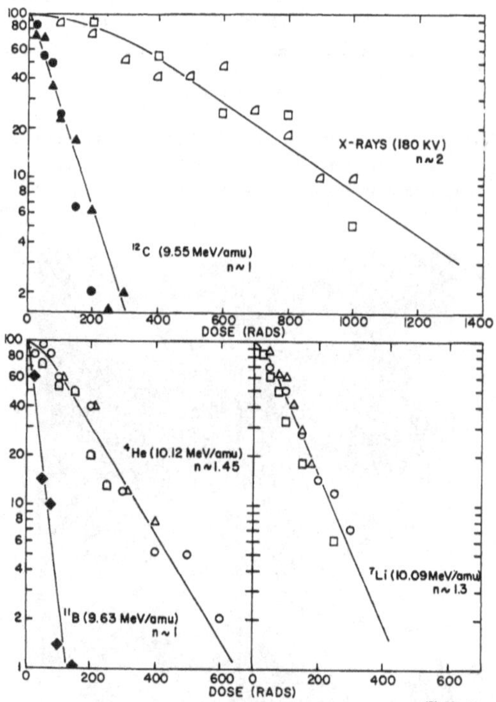

Fig. 3. Response of *T. confusum* eggs to heavy ion and X-ray irradiation.

still at a beginning stage. Using the facilities of the Heavy Ion Linear Accelerator (HILAC) at Lawrence Berkeley Laboratory, we have conducted some experiments to study the potential effects of heavy ions on biological development. Since the penetration power of heavy ions accelerated at the HILAC is limited, *Tribolium confusum* eggs, about 6 to 12 hours old at 23 °C, were chosen as experimental materials due to their small size. Hatchability dose curves for X-rays and heavy ions of different atomic number are shown in Fig. 3. A comparison of these curves indicate that there is a tendency for the initial shoulder of the dose-response curve to disappear gradually with increase in atomic number Z. The extrapolation number for X-rays, for example, is about 2, for lithium ions about 1.3, and for boron and carbon ions about 1. This result suggests that organisms can tolerate a certain amount of injuries induced by low-LET radiation without significant alteration in functional state, but cannot tolerate high-LET heavy ion injury. Evidently living animals doe not possess a mechanism which can keep their interior environment in balance after having been exposed to high-LET heavy ions. Experiments with other biological systems, e.g. mammalian cells and corn seeds, have given similar results, suggesting that heavy ions are extremely hazardous to biological systems [3—5], and that a single hit can cause some irreversible damage to the organism.

2.3. Gravity

The only environmental factor which has been maintained as a constant is the gravitational force to which all living organisms on earth have been genetically and physiologically adapted. Any change in this parameter, to a quantity either greater or less than "$1g$", produces certain effects in organisms. Deleterious changes in plants as well as animals under acceleration, or high g, have been observed. Inhibition of growth in wheat seedlings and young hamsters, for example, has been reported under high g [6, 7]. Reactions to the absence of gravity (i.e. weightlessness) in plants have been studied by many people using the clinostat, and a certain amount of information has consequently been accumulated [8, 9]. Very little, however, is known about responses of animals to weightlessness. It is difficult to study animals in the clinostat, which provides compensated gravitation through a uniform, horizontal, rotating motion, since the organism is required to be small in size, easily held in a fixed position in a simple environment, and must have a definite geotaxis. *Tribolium* admirably fits all these requirements; its pupae and adults are about 3 to 4 mm in size, it can survive in a simple flour-yeast medium, and adults have a negative geotaxis, i.e. beetles have a tendency to move to the opposite direction to gravitational force [10]. Geotaxic characteristics are very useful when one needs to know the minimum rotating speed that will provide conditions of "true" compensated gravity for the organism concerned. Some experiments have been done in our Laboratory using two 1π clinostats to study the responses of *Tribolium* to compensated gravity at different developmental stages. A full description of the 1π clinostat has been reported by Silver [11, 12]. When young eggs (less than 6 hours old at 30 °C) were subjected to gravity compensation during the entire egg stage, no detectable effect on the embryonic development was observed. Possibly the structure of the egg, consisting of a large quantity of yolk in which a number of cytoplasmic threads are distributed, has been modified through evolution so that it has become insensitive to changes in gravity. In fact, eggs are deposited randomly in the medium by adults and are constantly tumbled around by larvae and adults.

Since the size of *Tribolium* is limited, it is possible to keep the beetles in the clinostat through the entire life cycle. Recently we have exposed eggs as well as larvae and pupae to compensated gravity, and have subsequently observed some interesting effects. Pupal death and molting failures appeared when young larvae (2 to 3 weeks old at 30 °C) were rotated around a horizontal axis at a rate of 16 rev min^{-1} in our clinostat for a period of 30—40 days (Table 1). No such effects were observed in groups rotated at the same angular velocity for the same time, or in control groups kept stationary at $1g$ (Fig. 4). Pupal death and molting

Table 1
Development in the Flour Beetle (*T. confusum*) when exposed to Gravity Compensation

Condition	Total No. larvae used (2–3 wks old; 30°C)	Percentage developmental failure
Stationary	60	0%
Rotation, axis parallel to g	60	6.7%
Rotation, axis perpendicular to g	60	56.7%

failure are observed frequently when larvae or pupae have been X-irradiated, and appear as a consequence of disturbance to the endocrine system. Compensated gravitation may unbalance some of the physiological functions of *Tribolium*. Caution is needed in interpreting the present results, however, since some mechanical damage to *Tribolium* due to tumbling cannot be completely ruled out; more investigations are required.

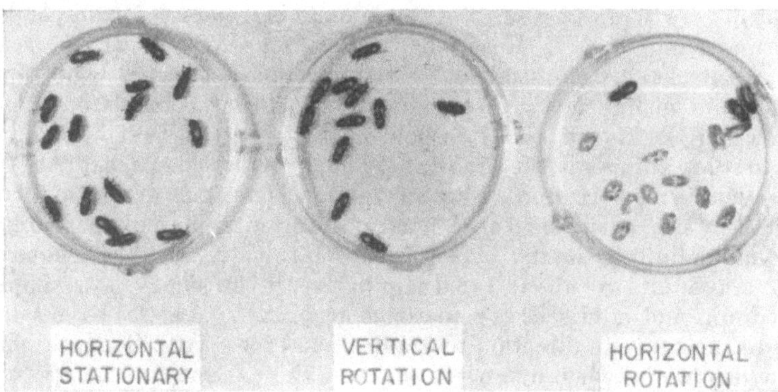

Fig. 4. Effect of compensated gravity on the development of *Tribolium*: left, *Tribolium* kept at stationary at $1g$; middle, *Tribolium* were rotated around vertical (g) axis; right, *Tribolium* developed under gravity compensation, i.e. rotation around horizontal axis, perpendicular to gravity. Size of the petri dish shown in the picture = 50 mm × 12 mm.

3. Interactions of more than one Environmental Factor on Living Organisms

How living organisms react under the action of combined environmental factors is a complicated but fascinating problem. Among the many possible combinations, the combined effects on pupal development of radiation, magnetic field, temperature, and oxygen concentration have been extensively studied by Amer [13]. It was found that magnetic field within a certain range of intensity has a "protective" effect on *Tribolium* pupae. The incidence of wing deformity in *Tribolium* beetles induced by high temperature (38 °C) and X-rays (1200 r) decreased with corresponding increase in magnetic field up to 8 kG. This "protective" effect, it was suggested, may be due to the reduction of the number of degrees of freedom of molecular motion under magnetic field. It became weaker in a higher magnetic field (10 kG); the reason for this is still unclear. The effect of magnetic field on X-irradiated developing pupae at various oxygen tensions was also investigated. At low oxygen concentration, no significant influence could be detected, but an enhancement was observed at high oxygen tension (Fig. 5). Since the hormones governing development and growth in this biological system are either steroids or precursors of steroids, it has been suggested that

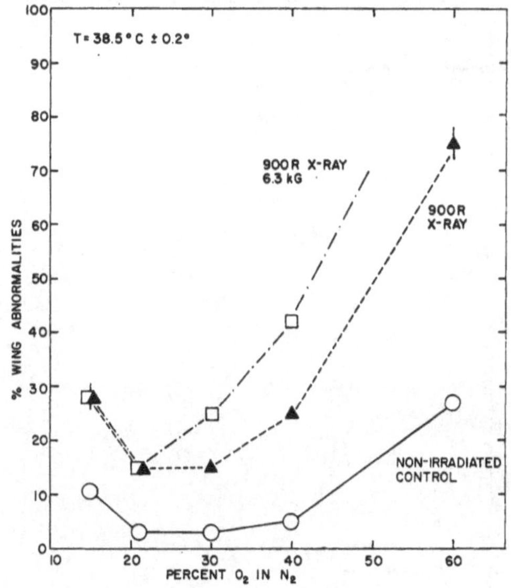

Fig. 5. Combined effects of magnetic fields, X-irradiation, ambient gas composition, and incubation at 38 °C upon wing abnormality.

oxidation of steroids may play a role in oxygen toxicity, and that magnetic fields may play an important part in influencing the rate of oxidation of unsaturated steroids.

Recently, we have focused our attention on the dynamic reactions of pupae to a wide range of temperature and radiation. With increase in one type of stress, it appears that the optimal range for another variable for *Tribolium* development decreases. When developing pupae received a dose of 1200 r, for example, the optimal range of temperature for their development was narrowed to 29—30 °C, as shown in Fig. 1. This optimum temperature range was further decreased as pupae were irradiated with 1600 r of X-rays. Results thus indicate that the capacity for homeostasis in the organism does not remain the same with various combinations of environmental factors. Synergism between radiation and temperatures located outside the optimum range is quite evident.

In order to obtain more quantitative information concerning this enhancement effect, we have designed and performed certain experiments. When pupae, aged from 18 to 24 hours old at 30 °C, were exposed to 5 °C at various intervals, the number of wing deformities increased with length of exposure time and reached

50% at 10 days exposure. Synergistic effect was observed when pupae were exposed to low temperature immediately following X-irradiation (Fig. 6). With increase in X-ray dose, the initial shoulder of the response curve gradually becomes smaller, and one would get an exponential curve with a moderately high dose of radiation. At varying X-ray doses, with all pupae exposed to 5 °C for the same interval (2 days), the amount of enhancement of X-ray effect with temperature was found to be dose dependent (Fig. 7).

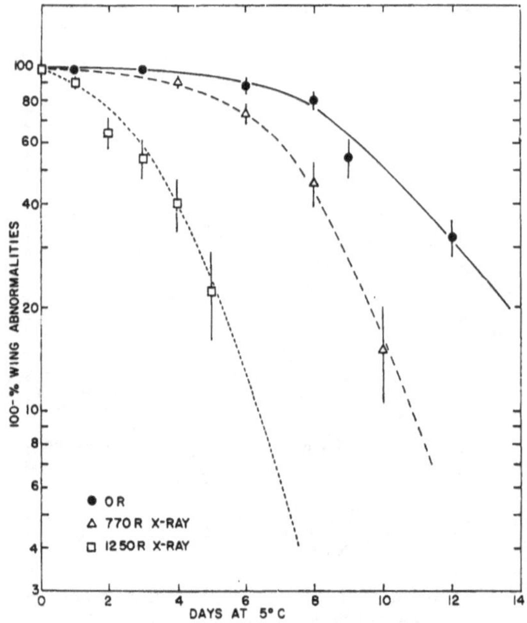

Fig. 6. Effect of low temperature (5 °C) on the wing development of *T. confusum* pupae irradiated with X-rays.

Experimental approaches have shown that the point of greatest interaction between the effects of environmental variables is not found near the optimum value of each of these variables. Rather it is near the extreme values which are the limits of tolerance of the organism in question for that environment.

Effects of weightlessness and radiation on the development of *Tribolium* pupae were also investigated in the Biosatellite 2 Flight Experiment. We observed that pupae which received a portion of their radiation exposure in orbit produced more wing abnormalities than controls kept at ground level or than controls irradiated at ground level and subsequently flown in space. There appears to be synergism between the effects of radiation and weightlessness on pupae development [14, 15]. In order to learn more about the effects of radiation and weightlessness, a clinostat combined with a cesium gamma-ray source has been built in our Laboratory. Using this 1π clinostat-irradiation system, we have found a qualitatively similar effect to that observed on Biosatellite 2 and reported in [16]. When young pupae (22—27 hours old at 30 °C) were horizontally rotated at a speed of 4 rev min^{-1} in the clinostat with the temperature constantly maintained at 38 ± 1 °C, and exposed to cesium gamma rays for several hours, an increase of about 4—9% in wing abnormalities was observed within a dose range of 2000 to 3000 r. This increase was not seen in groups of pupae vertically rotated or not rotated at all. A possible explanation for this synergistic effect between weightlessness and radiation might be that repair processes for radiation injury are less efficient under weightless conditions than at ground level.

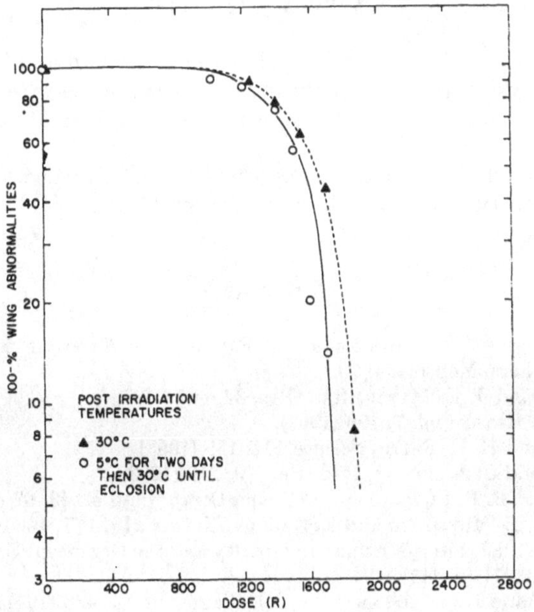

Fig. 7. Effect of low temperature (5 °C) treatment on the radiosensitivity of *T. confusum* pupae.

4. Summary and Conclusion

The results show that, for each environmental factor there is a limited range of values within which the organisms are able to best survive and that this optimum range of survival becomes smaller when additional stresses are imposed upon the organism. This has been shown for external temperature, oxygen content of the atmosphere, gravity compensation, and radiation. The effects of various environmental stresses, if the stresses are applied together, are not additive: the effects are nonlinear and synergistic when at least one of the environmental variables is near the limit of tolerance. Heavy-ion irradiation, which is present in primary cosmic rays, has been found to be extremely hazardous to living organisms, as irreversible injuries can be induced with a very small quantity of high-LET heavy ions.

On Biosatellite 2 it was shown that *Tribolium* pupae irradiated in space flight exhibited more wing abnormalities than ground controls. Later it was shown that gravity compensation, when combined with irradiation, can induce a similar effect at ground level. Life-support requirements for *Tribolium* are quite simple and it appears feasible to study a proliferating population of this organism for an extended period of time, perhaps on satellites such as the Skylab project. We expect to be able to determine definite synergistic effects of a weightless environment and heavy-ion radiation on the reproductive dynamics of the population as a whole and on the morphology of individuals.

Acknowledgments

The authors wish to thank Drs G. Welch and J. Lyman and Mr J. Howard for their help with physical measurements and dosimetry of heavy ions; Mrs L. Craise and Mrs B. Heinze for their technical assistance; and Miss D. Mundy for help in typing and preparation of this paper.

For the continuous interest and support of the US National Space and Aeronautics Administration and the Atomic Energy Commission, we are especially grateful.

References

[1] H. SELYE, Stress — The Physiology and Pathology of Exposure to Stress, Acto Inc., Medical Publishers, Montreal 1950.
[2] B. BUCKHOLD and J. V. SLATER, Rad. Res. **37**, 567 (1969).
[3] P. TODD, Rad. Res. Suppl. **7**, 196 (1967).
[4] H. J. CURTIS and H. H. SMITH, Science **141**, 158 (1963).
[5] J. V. SLATER and C. A. TOBIAS, Rad. Res. **19**, 219 (1963).
[6] S. W. GRAY and B. F. EDWARDS, J. Cell. and Comp. Physiol. **46**, 97 (1955).
[7] C. C. WUNDER, B. MILOJEVIC and L. EBERLY, Nature **210**, 177 (1966).
[8] S. A. GORDON and J. SHEN-MILLER, in: Gravity and the Organism, University of Chicago Press, Chicago 1971 (p. 415).
[9] H. M. CONRAD and K. YOKOYAMA, Physiol. Plantarum **24**, 426 (1971).
[10] T. PARK, Quart. Rev. Biol. **9**, 36 (1934).
[11] I. L. SILVER, Lawrence Berkeley Lab. Rep. No. 511 (1971).
[12] I. L. SILVER, Lawrence Berkeley Lab. Rep. No. 596 (January 1972).
[13] N. M. AMER, Ph. D. Thesis, University of California, Lawrence Radiation Lab. Rep. No. 16854 (1965).
[14] J. V. SLATER, B. BUCKHOLD and C. A. TOBIAS, BioScience **18**, 622 (1968).
[15] B. BUCKHOLD, J. V. SLATER, I. L. SILVER, T. YANG and C. A. TOBIAS, in: NASA SP-204, 1971 (p. 79).
[16] C. H. YANG, B. D. HEINZE, I. L. SILVER and C. A. TOBIAS, Lawrence Berkeley Lab. Rep. No. 596 (January 1972).

Life Sciences and Space Research XII — Akademie-Verlag, Berlin 1974

RETINAL CHANGE INDUCED IN THE PRIMATE (*Macaca mulatta*) BY OXYGEN NUCLEI RADIATION

C. H. BONNEY[a], F. N. BECKMAN[a] and D. M. HUNTER[b]

[a]USAF School of Aerospace Medicine, Brooks Air Force Base, Texas, USA
[b]Wilford Hall USAF Medical Center, Lackland Air Force Base, Texas, USA

Retinas of primates* (*Macaca mulatta*) were exposed to oxygen nuclei at the Bevatron, Berkeley, California. Color fundus photographs and fluorescein angiograms were taken of the retinas prior to irradiation and up to 5 weeks post exposure. Animals were sacrificed at post exposure intervals for histopathologic examination of the retinas. A series of animals were exposed to 200 kVp X-ray and examined on the same regime as the first series. The results showed a low rad equivalent dose for retinal damage as compared with the X-ray series, i.e., a high quality factor, and a marked compression of the latency between exposure and onset of the retinal pathology.

1. Introduction

When crew members of Apollo 11 as well as subsequent Apollo crews reported seeing "flashes of light" [1—3] verification was given to a prediction made more than a decade and a half earlier by Dr. Tobias that man in space might expect to see cosmic rays [4]. Disagreement exists as to the mechanism by which these particles interact within the eye to produce the visual event [5—7]. On a larger scale, the whole realm of high energy particle interaction with tissue is almost unknown [5, 8]. An extension of research supported by the National Aeronautics and Space Administration (NASA) at the School of Aerospace Medicine, Brooks Air Force Base, San Antonio, Texas, has been an effort to study the deleterious effects of high energy particles upon the retina of primates (*Macaca mulatta*). In cooperation with Dr. Tobias and his colleagues at the Lawrence Radiation Laboratory, Berkeley, California, a series of exposures have been made utilizing a beam of accelerated oxygen nuclei.

The retina provides a highly organized, accessible portion of the central nervous system with vascular and neural elements representing relationships found throughout the central nervous system. Most importantly, the retina may be viewed and photographed within the living animal. Additionally, visual function is of paramount importance to crew performance.

It was the purpose of this study (i) to determine the temporal relationship of the nature and extent of retinal damage; (ii) to examine the relative biological

* The animals involved in this study were maintained in accordance with the "Guide for Laboratory Animal Facilities and Care" published by the US National Academy of Sciences, National Research Council.

effectiveness of this form of radiation by comparing retinas exposed to heavy ion radiation with retinas exposed to the same dose in rads of X-irradiation; and (iii) to assess the visual acuity of exposed primates by using operant conditioning technics.

2. Methods

2.1. Oxygen Nuclei Series

Forty primates (*Macaca mulatta*) were exposed to a beam of oxygen nuclei. These animals were sedated with phencyclidine HCl so that a catheter could be introduced into a saphenous vein. Anesthesia was induced by administering sodium pentobarbital through the catheter. The eye to be exposed, the left eye, was dilated using ophthalmic phenylephrine HCl and atropine sulphate. The animals were placed in a visual stereotaxic instrument mounted on a stand having three degrees of freedom. A Zeiss fundus camera was mounted in front of the stand in order to photograph the retina. The stereotaxic instrument with the

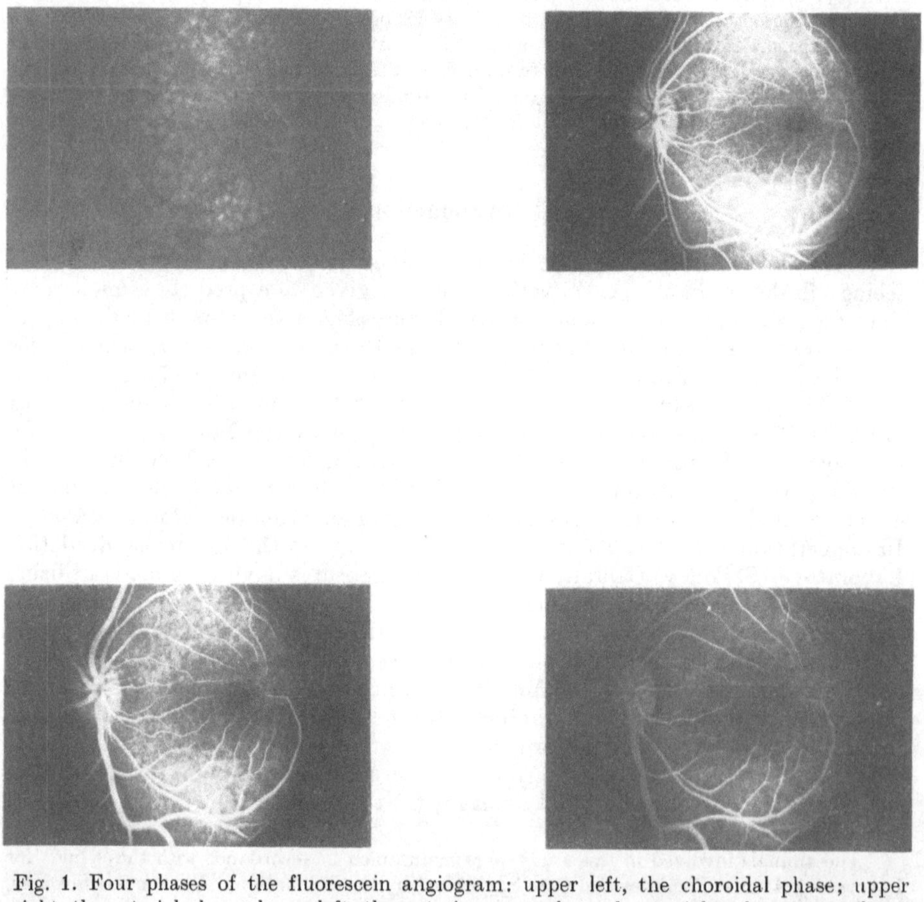

Fig. 1. Four phases of the fluorescein angiogram: upper left, the choroidal phase; upper right, the arterial phase; lower left, the anteriovenous phase; lower right, the venous phase.

primate was moved to an identical stand in front of the oxygen nuclei beam so that the beam axis was identical with the camera axis. Thus the beam struck the retinal surface which had been viewed by the camera. Irradiation was conducted with the eyelid closed.

A clinical ultrasonic unit was used to measure the lid-to-retina axis of the globe to be irradiated. These measurements were made through the eyelid. This measurement was used to adjust a water column in the beam path so that the Bragg peak of the beam fell 0.5 mm posterior to the inner surface of the retina. Prior to and immediately following exposure to the beam, fundus photography and fluorescein angiograms were taken. Three quarters of a milliliter of 10% sodium fluorescein was injected into the circulation through the saphenous vein catheter. These procedures were repeated at post-exposure intervals of 24 hours, and 1, 2 and 5 weeks (Fig. 1). At intervals of 24 hours, 2 and 5 weeks, animals were perfused through the abdominal aorta with glutaraldehyde; the eyes were removed and prepared for embedding in plastic. Sections were cut at 1 μm and stained with toluidine blue [9]. The number of primates exposed, the flux of particles, the retinal irradiation in rads, and the number of animals perfused are given in Table 1.

Table 1

Oxygen Nuclei Exposure of Primate Retinas

Number of animals	Exposure		Post-exposure Sacrifice				
	Flux (particles cm^{-2})	Retinal dose (rads)	24 hrs	1 week	2 weeks	5 weeks	Long term*
9	1.3×10^7	170	3		3	3	
9	3.9×10^7	500	3		3	3	
3	5.5×10^7	700					3
4	7.7×10^7	1000				2	2
6	1.5×10^8	2000			3	3	
4	2.3×10^8	3000				2	2
5	3.1×10^8	4100				2	3

* Animals maintained for further study.

2.2. X-ray Series

For comparison, two series of primates were prepared as described above and exposed to 200 kVp X-rays. The exposures were made through the eyelids as were the heavy ion exposures. One series of five animals was exposed at a dose rate of 65 rads per minute for a total dose of 3000 rads. The second series was exposed at the same rate for a total dose of 4150 rads. Dosimetry was based on ion chamber measurements in both plastic and TLD dosimeters in an Alderson primate phantom. Color fundus photographs and fluorescein angiograms were also taken of these animals at post-exposure intervals of 24 hours, 1, 2 and 5 weeks.

2.3. Visual Acuity

Four primates trained to respond to the Landolt-C were exposed to provide an assessment of visual acuity changes [10]. Two animals were exposed to 2.3×10^8 particles cm^{-2} and two to 7.7×10^7 particles cm^{-2} rads of oxygen nuclei.

These animals were anesthetized as described and exposed to oxygen nuclei irradiation. Fundus photographs and fluorescein angiograms were taken of these animals immediately before and after irradiation exposure. Visual acuities were measured following exposure for comparison with published radiation induced acuity changes [11, 12, 14, 22].

3. Results

3.1. Oxygen Nuclei Series: Fundus Photography and Fluorescein Angiograms

The earliest funduscopic observation was of small, discrete retinal hemorrhages seen first at the 1.3×10^7 particles cm^{-2} dose 24 hours after exposure (Fig. 2). These lesions were small transient hemorrhages of the order of 0.1 mm.

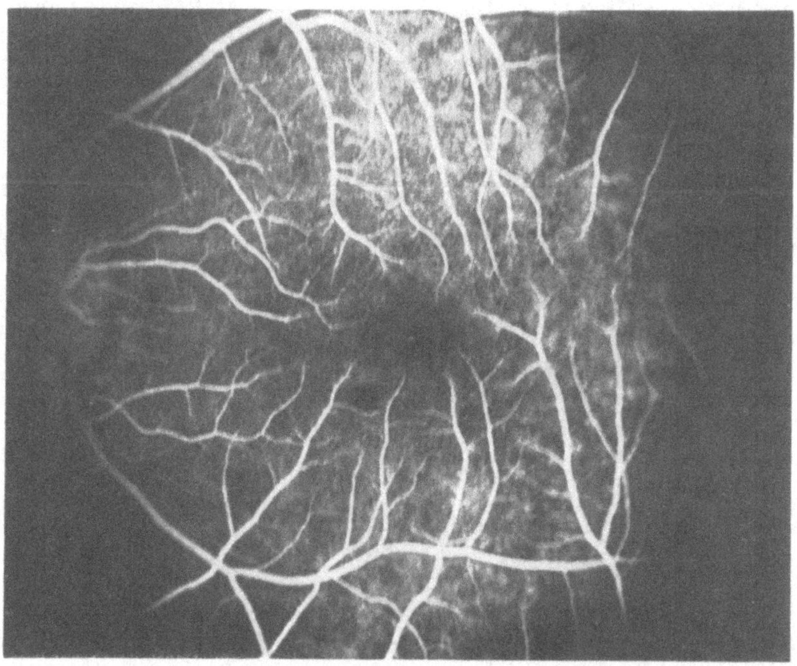

Fig. 2. A retinal hemorrhage on the inferior border of the macula.

7.7×10^7 particles cm^{-2}. The most consistent early change has been dye leakage seen in the angiograms at 24 hours following irradiations of 7.7×10^7 particles cm^{-2} or more (Fig. 3). The leaks were small and circumscribed along both the arteries and veins. These leaks were not visible on the color fundus photograph (Fig. 4) although there were increased highlights from the vitreal-retinal interface. It was not until the one-weak funduscopic examination that a lesion was observed at the site of radiation and was characterized by hemorrhage, loss of retinal transparency and ghosting of capillaries (Figs. 5a, b). This had the clinical appearance of a cotton wool patch which was associated with ischemic necrosis in the ganglion

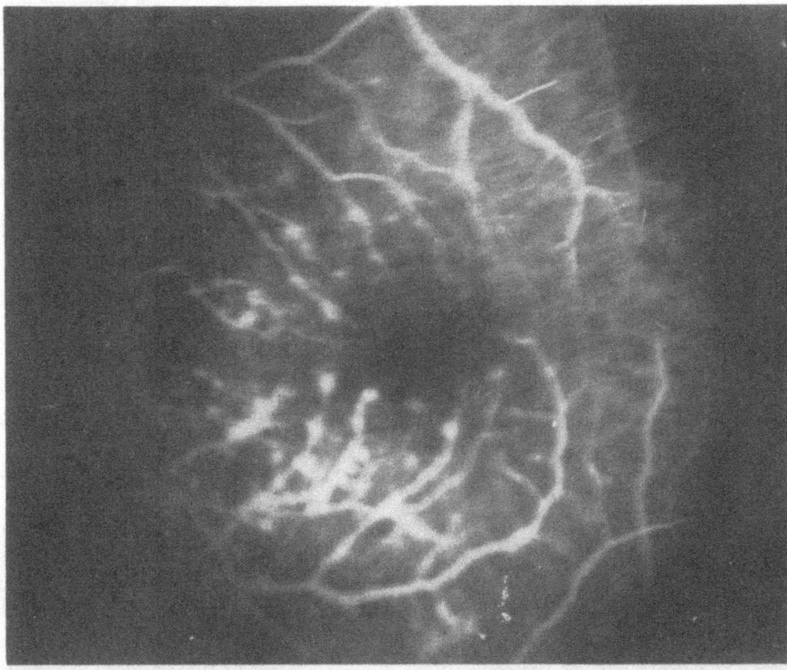

Fig. 3. Dye leakage 24 hours after exposure from both the inferior and superior arcade of retinal vessels. Exposure was 7.7×10^7 particles cm^{-2}.

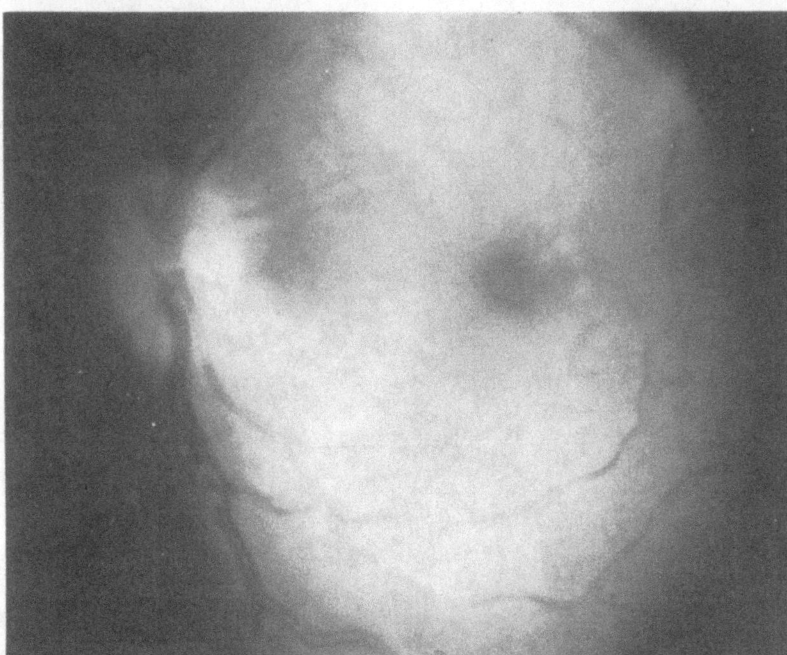

Fig. 4. Retina with increase in highlights from the vitreal-retinal interface.

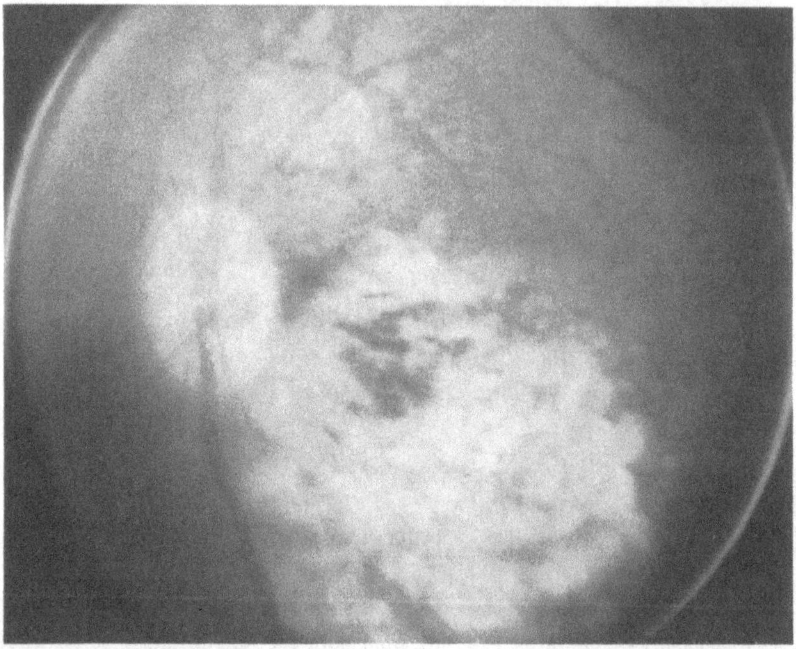

Fig. 5a. Retina lesion with hemorrhages and loss of retinal transparency.

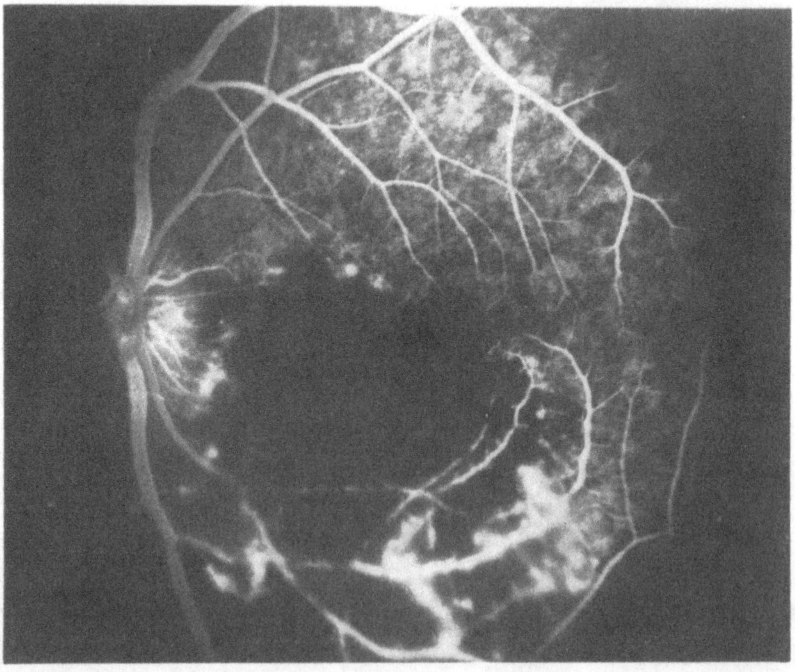

Fig. 5b. The fluorescein angiogram.

cell layer. In the angiogram there was a loss of capillary perfusion in the central aspect of the lesion. At the periphery leakage of the dye continued. At two weeks the picture was essentially that seen at one week except that dye leakage was diminished. Following an interval of five weeks the retinal opacities had been resolved with increased pigmentation visible on the fundus photographs. The

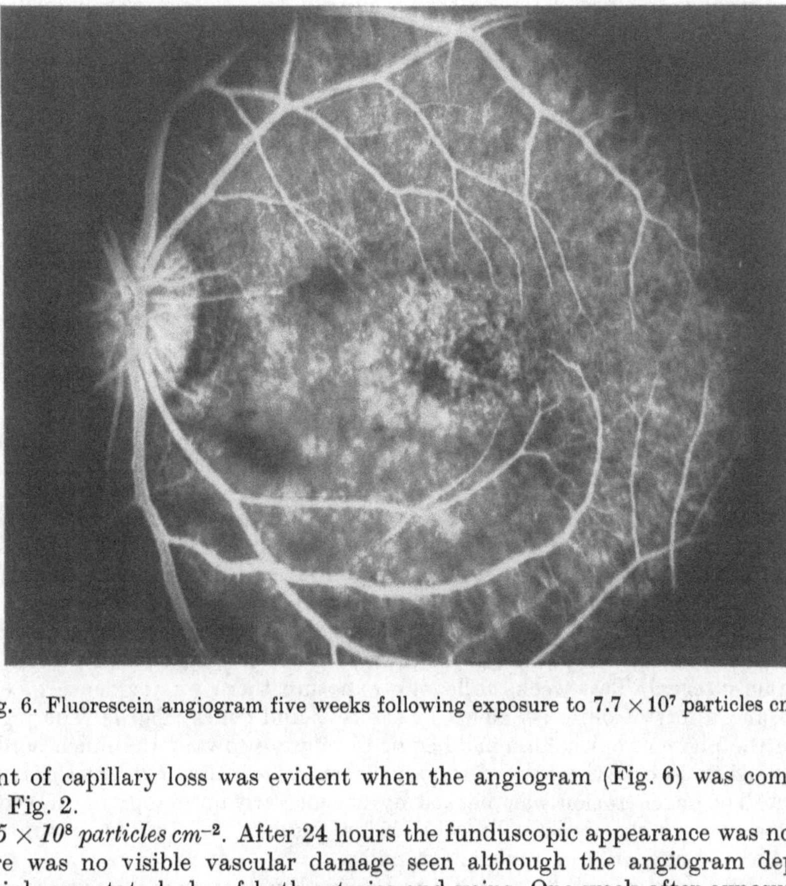

Fig. 6. Fluorescein angiogram five weeks following exposure to 7.7×10^7 particles cm^{-2}.

extent of capillary loss was evident when the angiogram (Fig. 6) was compared with Fig. 2.

1.5×10^8 particles cm^{-2}. After 24 hours the funduscopic appearance was normal. There was no visible vascular damage seen although the angiogram depicted multiple punctate leaks of both arteries and veins. One week after exposure the severity of the lesion was somewhat greater than that seen following the exposure to 7.7×10^7 particles cm^{-2}. Hemorrhages were diffuse over the lesion and appeared to be in the nerve fiber layer. Congested capillaries seen in the color photographs were shown by the angiograms to be non-functional. Five weeks after exposure there was a return of retinal transparency generally. The periphery of the lesion showed evidence of unresolved retinal edema. The angiographic record showed the same loss of capillaries throughout the lesion as previously described. Again the capillary loss was sharply demarcated from the functional vessels. There was some mottling of the retinal pigment epithelium.

2.3×10^8 particles cm^{-2}. Twenty four hours following exposure the funduscopic picture was retinal ischemia with retinal edema and loss of transparency. Hemorrhages in the nerve fiber layer were seen. The angiograms demonstrated non-

functional capillaries within the lesion. Dye leakage was seen to continue from capillaries at the periphery of the lesion. At one week the lesion exhibited edema and resultant loss of retinal transparency. Hemorrhages were still present. Angiographic records revealed essentially the same picture as seen at the one week interval. Two weeks following exposure there was some resolution of the retinal edema. Hemorrhages continued to occur. The angiograms still showed dye leakage from vessel peripheral to the lesion but the leakage was judged to be much reduced. Return of retinal transparency had occurred at the 5 week interval. The angiograms continued to show the loss of functional vessels within the irradiated portion of the retina. Retinal pigment epithelial mottling was again observed.

3.1×10^8 particles cm^{-2}. Following this level of irradiation a very pronounced lesion was present after 24 hours. Retinal edema and opacity was severe, with diffuse hemorrhages seen throughout the retinal area exposed to the oxygen ions. The course of both the color fundus photographs and the angiograms parallels that of the lesions described above at lower fluxes and greater intervals after exposure, i.e. two-week lesions. Resolution of retinal edema following this exposure was seen five weeks after exposure. The angiograms continued to show a discrete border to the lesion and loss of capillaries within the lesion. A greater degree of mottling of the pigment epithelium was seen than at this same interval following lower dosages.

3.2. Oxygen Nuclei Series: Histopathology

Histopathological changes were observed first at 3.9×10^7 particles cm^{-2}. Changes were produced at one week in the cytoarchitecture of the outer segments of the rods and cones. There was a loss of the parallel organization of these structures as well as loss of photopigment disks from the outer segments.

7.7×10^7 particles cm^{-2}. Alterations within the cytoarchitecture of the retina were more pronounced after 24 hours than at lower exposure levels evaluated at the same interval. Two weeks following exposure there was evidence of cytoid bodies and a loss of outer segments of the rods and cones (Fig. 7). The pigment within the pigment epithelium had begun to migrate toward the inner border of the cell. At five weeks there was disorganization of the retina with areas of necrosis present. The degeneration was marked by vacuoles within cytoplasm, karyolysis, and loss of staining of the cells. The retinal pigment epithelium demonstrates clumping and migration of pigment granules.

1.5×10^8 particles cm^{-2}. The earliest histological evaluation of tissue made at this exposure was at the two-week interval. The retina again showed a generalized necrotic retinitis. The pigment epithelium showed vacuolization, pigment migration, and loss of cellular architecture. At five weeks the picture of the retina was still that of a generalized retinal necrosis with loss of identifiable fiber or cellular layers of the retina.

2.3×10^8 particles cm^{-2}. Twenty four hours following exposure the retina showed a generalized necrotic retinitis. Loss of both cellular and fiber components of the retina were evident. Pigment clumping within the pigment epithelial cells was seen. A retinal detachment was present in the macula. Two weeks after exposure the more involved areas demonstrated loss of the outer segments. The plexiform layers could not be clearly distinguished and a decrease in the number of cells in both the inner nuclear and ganglionic cell layers was evident (Fig. 8). Five weeks

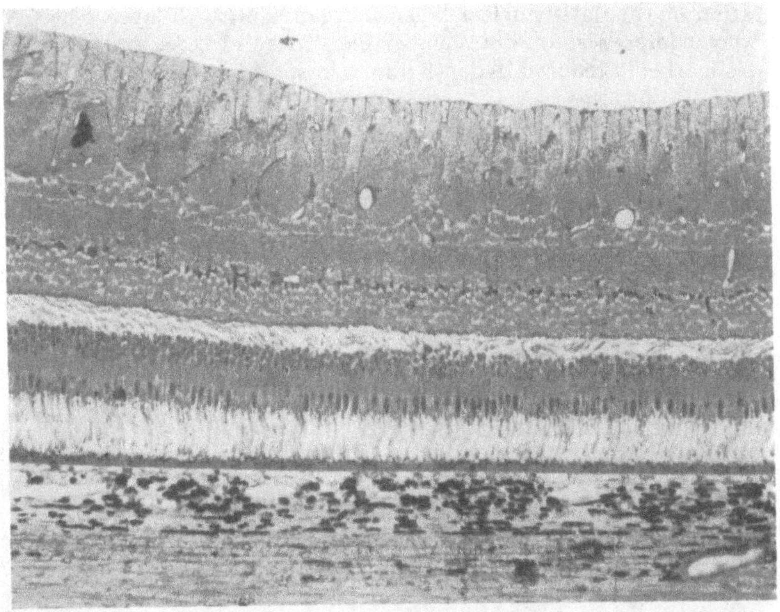

Fig. 7. Retina showing loss of outer segments, edema, and cytoid formations.

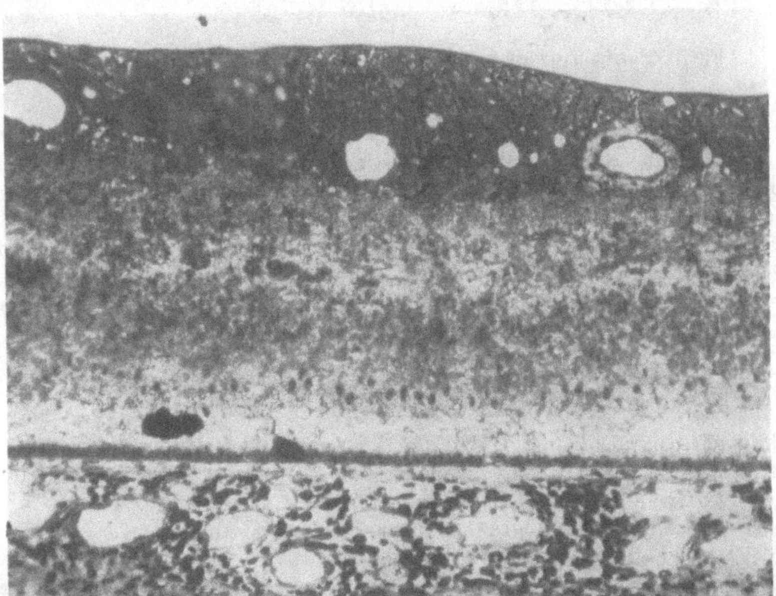

Fig. 8. Retina showing a generalized necrosis with loss of identifiable nuclear and fiber layers.

after exposure there was a generalized necrosis of the rods and cones, with vacuolar degeneration of the outer nuclear layer and ganglionic cell layer. The plexiform layers were compressed or not identifiable. These changes produced a retina which was markedly reduced in depth due to loss of fiber and cellular constituents (Fig. 9).

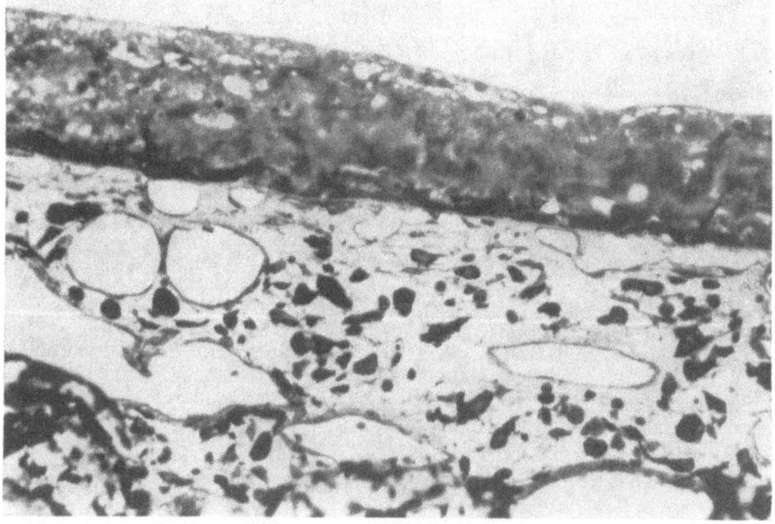

Fig. 9. Retina showing a generalized loss of nuclear and fiber elements.

3.1 × 10⁸ particles cm⁻². Following doses of 4000 rads or more there was massive destruction of the retina at 24 hours with almost complete loss of tissue by the end of two weeks. There was loss of the outer segments in association with edema of the outer nuclear layer. The other layers of the retina had been destroyed. There were retinal detachments present with evidence of serous accumulations between the pigment epithelium and remnants of the rods and cones. Rounding of the pigment epithelial cells had occurred. Glial activity was evident in the inner aspect of the retinal remnant.

3.3. Oxygen Nuclei Series: Visual Acuity

Following an exposure of 2.3×10^8 particles cm^{-2}, the acuity of two of the trained primates was 20/200. Of these animals, one showed an improvement to 20/100 two weeks after exposure. This animal remained at the 20/100 level for a week and then showed 20/30 and 20/20 acuities. The second animal exposed to 2.3×10^8 particles cm^{-2} had no measurable visual acuity one week after exposure. The unexposed eye of the second animal was tested and found to have 20/20 or 20/30 visual acuity. Fundus photography of the animals which received 2.3×10^8 particles cm^{-2} revealed massive retinal hemorrhages in the retina with large infarcted areas. Some variation in acuity was seen in both animals exposed to 7.7×10^7 particles cm^{-2}. This variation was present for the first two weeks after exposure. One animal showed an acuity in the exposed eye of 20/50 12 days after exposure before returning to the 20/20 or 20/30 level. The second animal showed a fluctuation between acuities of 20/100 and 20/30 between the tenth and sixtieth days after exposure.

3.4. X-ray Series: Fundus Photography and Fluorescein Angiography

In contrast to the oxygen nuclei irradiation, the X-irradiation showed no evidence of altered capillary permeability. The principal lesions were in the structures of the anterior segment. These changes were moist desquamation of the lids, iritis, and a panconjunctivitis.

3.5. X-ray Series: Histopathology

Histologically, the retinal changes in both series were slight pyknosis of the rod nuclei with some vacuolization and cloudy swelling of bipolar and ganglion cells.

4. Discussion

Radiation retinopathy has been reported in both clinical and experimental literature as primarily a circulatory lesion with other retinal changes being secondary to the vascular damage. Manifestations of the circulatory lesions are hemorrhages, occlusion of small vessels, microinfarction, telangiectasis, microaneurysms, leakage of fluorescein, and retinal edema. As the retinal syndrome progresses, a choroiditis may develop with exudates, pigmentary changes, occlusion of vessels [11—19], neovascularization, and finally retinal atrophy. The radiation threshold for the production of a chronic retinopathy has been placed at approximately 2000 rads [14, 20, 21].

An important aspect in the manifestation of radiation demage in the retina is that of latency. Latencies of three months to three years between exposure and the onset of retinal changes has been seen clinically in patients given X- or gamma rays [12, 14, 16, 18, 22]. The relationship between radiation dosage and retinal changes is expressed as a change in latency, i.e. the greater the dose, the shorter the latency.

The early changes within the monkeys were retinal hemorrhages and altered capillary permeability, indicating that the oxygen nuclei irradiation, like other forms of irradiation, produces changes first in the retinal vasculature.

At 5.5×10^7 particles cm^{-2} and below there is no evidence of changes in the angiograms. The histopathological evidence at 3.9×10^7 particles cm^{-2} indicates that some cellular alterations in the outer segments had occurred. A longer post-exposure following of these animals might have revealed changes following a latent period. The evidence available indicates, however, that these animals were near the threshold for damage.

As the dosage was increased the histological lesions appeared more quickly, except for the tissue examined at two weeks after 2.3×10^8 particles cm^{-2}, but the possibility must be considered that in this experiment the narrow Bragg peak fell in the vitreous.

The frank necrotic retinitis seen at 24 hours following 3.1×10^8 particles cm^{-2} is identical in clinical appearance with that reported 9 to 22 months after application of cobalt-60 plaques directly on the globe giving dosages of 20000 rads to the retina. The comparison between the findings with 4100 rad of X-rays and oxygen nuclei (Fig. 8) reinforces this striking difference in radiation quality factor.

The irradiation with oxygen nuclei demonstrated marked compression of the time scale on which events of the retinal radiation syndrome occur and at a dosage well below that necessary to produce comparable changes with cobalt-60. Thus, from both the reports of clinical cases using cobalt-60 and our comparison between 4100 rads of X-ray and oxygen nuclei, it is seen that the oxygen ion has a high quality factor. Thix factor may be of the order of 6 or 7.

Alterations in the acuities of all animals in the first 14 days after exposure is attributable to a residual cycloplegic effect of the atropine sulfate administered before exposure for its mydriatic effect. Acuities of 20/30 were taken as normal and were not interpreted as radiation-induced changes in acuity. Those animals exposed to 1000 rad will be maintained and tested for evidence of latent changes, and this will be reported later.

References

[1] T. F. BUDINGER, H. BICHSEL and C. A. TOBIAS, Science **172**, 868 (1971).
[2] I. R. MCAULAY, Nature **232**, 421 (1971).
[3] G. L. WICK, Science **175**, 615 (1972).
[4] C. A. TOBIAS, J. Aviation Med. **23**, 345 (1952).
[5] G. M. COMSTOCK et al., Science **172**, 154 (1971).
[6] G. G. FAZIO, J. V. JELLEY and W. N. CHARMAN, Nature **228**, 260 (1970).
[7] D. E. PHILPOTT, R. CORBETT, S. BLACK and C. TURNBILL, Proc. 30th Ann. Conf. Electron Microscopy Soc. Amer., Los Angeles 1972.
[8] W. N. CHARMAN, J. A. DENNIS, G. G. FAZIO and J. V. JELLEY, Nature **230**, 522 (1971).
[9] P. S. COOGAN and F. MORRIS, SAM TR-69-53, Sept. 1969.
[10] E. S. GRAHAM, D. N. FARRER, G. H. CROOK and P. V. GARCIA, Behav. Res. Meth. and Instru. **2**, 301 (1970).
[11] W. A. BEDFORD, C. BEDOTTO and P. A. MACFAUL, Br. J. Ophthalmol. **54**, 505 (1970).
[12] P. H. Y. CHEE, Am. J. Ophthalmol. **66**, 860 (1968).
[13] P. CIBIS and D. BROWN, Am. J. Ophthalmol. **40**, 84 (1955).
[14] A. DE SCHRYVER, L. WACHTMEISTER and I. BÅRYD, Acta Radiol. **10**, 193 (1971).
[15] M. J. HOGAN and L. E. ZIMMERMAN, Ophthalmic Pathology, 2d edn, W. B. Saunders Co., Philadelphia 1962.
[16] G. M. HOWARD, Arch. Ophthalmol. **76**, 7 (1966).
[17] R. K. MARMUR and N. A. MANTURO, Radiobiologica **6**, 431 (1966).
[18] P. A. MACFAUL and M. A. BEDFORD, Brit. J. Ophthalmol. **54**, 237 (1970).
[19] F. W. NEWELL et al., Am J. Ophthalmol. **50**, 1215 (1960).
[20] P. A. CIBIS, W. K. NOELL and B. EICHEL, SAM TR-55-41, 1955.
[21] D. R. LUCAS, Int. J. Radiat. Biol. **5**, 345 (1962).
[22] M. PERRERS-TAYLOR, D. BRINKLEY and T. REYNOLDS, Acta Radiol. **3**, 431 (1965).

Life Sciences and Space Research XII — Akademie-Verlag, Berlin 1974

THE BIOSTACK EXPERIMENTS I AND II ABOARD APOLLO 16 AND 17

H. BÜCKER

Arbeitsgruppe für biophysikalische Weltraumforschung, Universität Frankfurt, Frankfurt/Main, FRG

The concept of the Biostack experiment has become practicable through European scientific collaboration and with help of NASA. The objectives of this experiment flown aboard Apollo 16 and 17 are to study the biological effects of individual heavy cosmic particles of high-energy loss (HZE) not available on earth; to study the influence of additional space flight factors; to get some knowledge on the mechanism by which HZE particles damage biological materials; to get information on the spectrum of charge and energy of the cosmic ions in the spacecraft; to estimate the radiation hazards for man in space.

For this purpose the Biostack experiment comprises a widespread spectrum of biological objects, and various radiobiological end-points are under investigation. Bacterial spores, protozoa cysts, plant seeds, shrimp eggs, and insect eggs were included in the Biostack experiment packages together with different physical radiation detectors (nuclear emulsions, plastics, AgCl crystals, and LiF thermoluminescence dosimeters). By using special arrangements of biological objects and physical track detectors, individual evaluation of tracks was obtained allowing the identification of each penetrating particle in relation to the possible biological effects on its path. The response of the different biological objects to space flight and HZE ions bombardment was of different degree, presumably depending on the ability of the organism to replace the cells damaged by a hit. The results help to estimate the radiation hazard for astronauts during space missions of long duration.

The objective of the Biostack experiments is to study the combined action of individual heavy nuclei of the cosmic radiation and space flight factors on biological systems in a state of rest. The Biostack experiment package contains a series of monolayers of selected biological objects in a fixed position, with each layer sandwiched between physical track detectors (Fig. 1). By using this arrangement, individual evaluation of tracks is obtained, allowing the identification of each penetrating particle in relation to the possible biological effects of its path. The heavy (high atomic number) hich energy particles that have high energy loss on their trajectory through material (HZE particles) are of special interest. Such particles are not available on earth. They are found in the spectrum of the galactic cosmic radiation. Knowledge of this radiation outside the earth's magnetic field is still very small.

Therefore it was of great importance to make use of the two last Apollo flights, since there will be no other mission into deep space for a considerable time. Detailed information on particle flux and the spectra of energy loss and charge is of essential importance for estimating the biological hazard of space missions.

The exposure of a Biostack experiment in the command modules of Apollo 16 and Apollo 17 and subsequent evaluation of the HZE particles encountered, together with their individual biological effects, had the following advantages: (i) data were made available on the biological effects of HZE particles, which contribute to basic research in radiation biology; (ii) the results of the Biostack

Fig. 1. The Biostack experiment package: left, container; right, stack consisting of biological objects in monolayers and physical detectors.

experiments may be of essential help estimating the radiation hazard in manned space missions; (iii) since a space ship can never be shielded against heavy primaries, they have to be considered in the evaluation of each biological space experiment in which other factors such as weightlessness are investigated. The Biostack data yield a basis also for these estimations.

The heavy nuclear component of the galactic cosmic radiation was discovered in 1948, rather late in the history of cosmic-ray research. Radiobiologists soon realized, however, that this new type of ionizing radiation represents also a new type of factor acting on living matter when exposed to it. Direct experimental evidence of the effectiveness of individual heavy particles of cosmic radiation was presented by Chase [1] in describing greying of hair in the coat of balloon-borne black mice. Eugster [2] demonstrated cellular destruction by single hits of heavy ions on *Artemia salina* eggs, and Brustad [3] on maize embryos. Brain injury has been studied by Yagoda et al. [4] in balloon-borne mice, and more recently by Haymaker et al. [5] in balloon-borne monkeys. The problem of hazards from primary cosmic radiation during space flight has been given increased attention since the astronauts of the Apollo 11 and following missions experienced light flashes as a visual phenomenon. Although this phenomenon was already predicted by Tobias in 1952 [6], attention was not given to the problem until the astronauts actually experienced the light flashes in space. In 1970, when a suitable beam was available at Berkeley, Tobias and his team reproduced the eye flash phenomena with these ions and thus showed that the light flashes seen by the astronauts were caused by heavy cosmic particles [7].

The very high local concentration of absorbed energy produced by an HZE particle may cause serious biological effects when cells of the central nervous system are concerned, because complete cells can be destroyed by such a microbeam and they cannot regenerate. From the physical data of a particle, in particular its energy loss, it is possible to calculate the absorbed dose. It will be possible to relate this value to dose-effect curves of X-rays or gamma-rays, but this will *not* lead to accurate estimation of the biological effects, because the application of the dose is quite different in the two cases, above all because the time is not comparable. The very high local concentration of absorbed energy, a phenomenon which is properly described by the term "microbeam" [8] is applied in a very short time. Including the time of energy transfer and energy deposition in biological matter, the application time of an HZE particle is less than 10^{-6} second, while in the usual irradiation experiments it is in the range of seconds, minutes or even longer. The only way to find out the biological effects of HZE particles is to conduct experiments with them.

A sophisticated method must be used, in order to localize precisely the trajectory of a particle relative to the biological objects, and to correlate the physical data of the particle with the observed biological effects along its path. In the Biostack experiment, special methods were developed for this purpose [9]. Selected biological objects and physical track detectors were used in order to achieve optimum information.

The following considerations were taken into account in the selection of the biological objects: (i) they had to endure fix arrangement between track detectors, e.g. embedding in PVA (polyvinyl alcohol); (ii) they should comprise a variety of biological specimens (such as plant seeds, animal eggs and bacterial spores) to allow evaluation of radiation effects on different levels of biologic organization; (iii) biological objects with different radiation sensitivity (as known from radiobiological experiments with X-rays, gamma-rays or electrons) should be used, since there is insufficient experience on the biological effectiveness of individual HZE particles, and these cannot be compared with common radiation sources.

Table 1 gives the biological objects which were exposed in at least one of the two space flight Biostack experiments, together with the effects under investigation and the individual investigators. The investigations performed in the Biostack II experiment flown aboard Apollo 17 were mainly based on the results obtained from analysis of Biostack I, flown with the Apollo 16 mission. Physical track detectors were selected to allow identification of the biological objects hit, localization of the area hit in the biological object, comprehensive information on the energy loss of the various HZE particles. Table 2 gives the radiation detectors used in the Biostack, together with their characteristics and the responsible investigators. All detectors, except the lithium fluoride, are track detectors.

Detailed controls were made in parallel with the space flight experiments with Biostack. For each space flight experiment, four identical Biostacks were built. In each case, three units were delivered to NASA: one prime flight unit, one backup, in case of damage to the prime flight unit, one ground control to remain in

Table 1

Biological Effects investigated in the Biostack Experiments

Biological system: General	Specific	Effects under Investigation	Investigators	Organization
Bacterial spores	*Bac. subtilis* 2 strains of different radiation sensitivity	Inactivation of germination and outgrowth Induction of abnormal cells	G. Horneck	University of Frankfurt, FRG
Protozoal cysts	*Colpoda cucullus*	Inactivation Modification of growth	H. Planel, J. P. Soleilhavoup	University of Toulouse, France
Plant seeds	*Arabidopsis thaliana*	Influence on germination, growth, and development Mutation induction	E. Reinholz	MPI für Biophysik, Frankfurt, FRG
	radicules of *Vicia faba*	Decrease of growth Nuclear damage	W. Scheuermann	University of Hanover, FRG
Crustacean eggs	*Artemia salina*	Induction of anomalies Influence on development	W. Rüther, E. H. Graul	University of Marburg, FRG
		Decrease of fertility	H. Planel, J. P. Soleilhavoup	University of Toulouse, France
Insect eggs	*Tribolium castaneaum* *Carausius morosus*	Induction of anomalies Influence on development Induction of histological anomalies	W. Rüther	University of Marburg, FRG

Table 2
Radiation Detectors in the Biostack Experiments

Cosmic Radiation Component	Detector	Range of Information on Z and LET	Threshold	Tissue Equivalence	Background Noise	Time Assignment	Investigators	Organization
Ions	Nuclear emulsion K2, K5	Very broad	No	No	High	No	P. Cüer, R. Kaiser, J. P. Massué, R. Pfohl	LPC-CRN Strasbourg
	Plastics: Cellulose nitrate, Polycarbonate	Medium	Yes	Yes	Low	No	W. Enge, O. C. Allkofer, K. Bartholomä, R. Beaujean, K. Fukui W. Heinrich	University of Kiel
							E. V. Benton	University of San Francisco
	AgCl-crystals (dark)	Narrow	Yes	No	No	No	E. Schopper, G. Henig,	University of Frankfurt
	AgCl-crystals (illuminated)	Broad	Yes	No	Medium to low	Yes	J. U. Schott	
γ-rays X-rays Protons	LiF (thermoluminescence dosimeter)		Integrating dosimeter				H. François, G. Portal	CEA, Paris

Houston. One laboratory control unit was kept in Frankfurt. Since the prime flight unit was flown in both missions the backup served as ground control and the two remaining units were available for further investigation. For Biostack I, a balloon flight at Fort Churchill, Canada, was conducted with one of the remaining units, while the other served as the relevant control. For Biostack II, one of the remaining units was irradiated at Berkeley at the Bevatron with carbon and oxygen ions, the other was subjected to vibration, acceleration and shock at the Centre National d'Etudes Spatiales (CNES) in Toulouse, France.

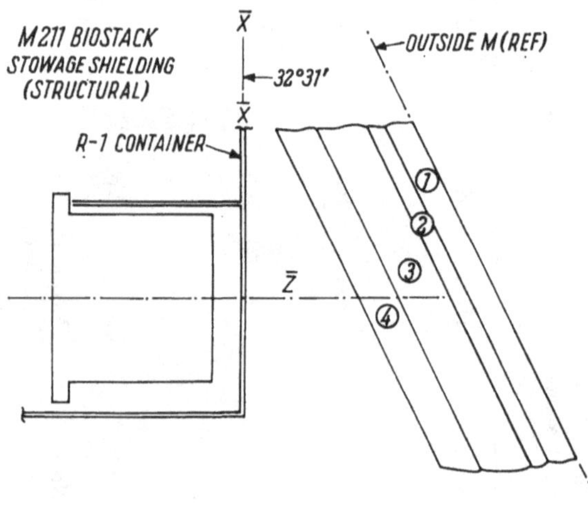

Fig. 2. Scheme of the position of the Biostack experiments in the Apollo command modules of Apollo 16 and Apollo 17 mission respectively (1 inch = 2.54 cm; 1 lb/ft^3 = 0.0160 kg cm^{-3}).

1. ablator, 1.78 cm thick	absorption	0.914 g cm^{-2}
2. steel honeycomb, 0.2 mm thick	,,	0.319 g cm^{-2}
3. insulation, 3.175 cm thick	,,	0.305 g cm^{-2}
4. aluminium honeycomb, 0.5 mm thick	,,	0.139 g cm^{-2}
wall of R 1 container, assumed as aluminium, 2.5 mm thick	,,	0.692 g cm^{-2}
	total absorption	2.369 g cm^{-2}

The Biostack I experiment was launched with Apollo 16 on 16 April 1972. Splash down into the Pacific was on 28 April. The total mission time was 266 hours. The temperature in the command module during the mission ranged mainly between 17 °C and 23 °C and the limits of 14 °C and 28 °C were never exceeded. The flight data of the Apollo 17 mission, with which the Biostack II was flown, were quite similar to those of Apollo 16. The total mission time of Apollo 17 was 304 hours. In both missions the Biostack experiment was placed in the R 1 compartment of the Apollo command module. Its position relative to the wall of the command module is shown in Fig. 2.

The approximate absorption in the four different layers of the wall of the command module was about 2.4 g cm^{-2}. The bottom of the Biostack container was aluminium 3.00 mm thick with absorption 0.84 g cm^{-2}. Since the software of the Biostack itself absorbs radiation, there is a decrease of radiation from outside to inside. Fig. 3 shows some data of the physical evaluation of Biostack I as a function of absorption. From this it is evident that the flux of efficient HZE particles behind an absorption screen of 20 g cm^{-2} is still half of the total flux encountered in the mission of Apollo 16. This result demonstrates that it is impossible to shield a space vehicle against HZE particles.

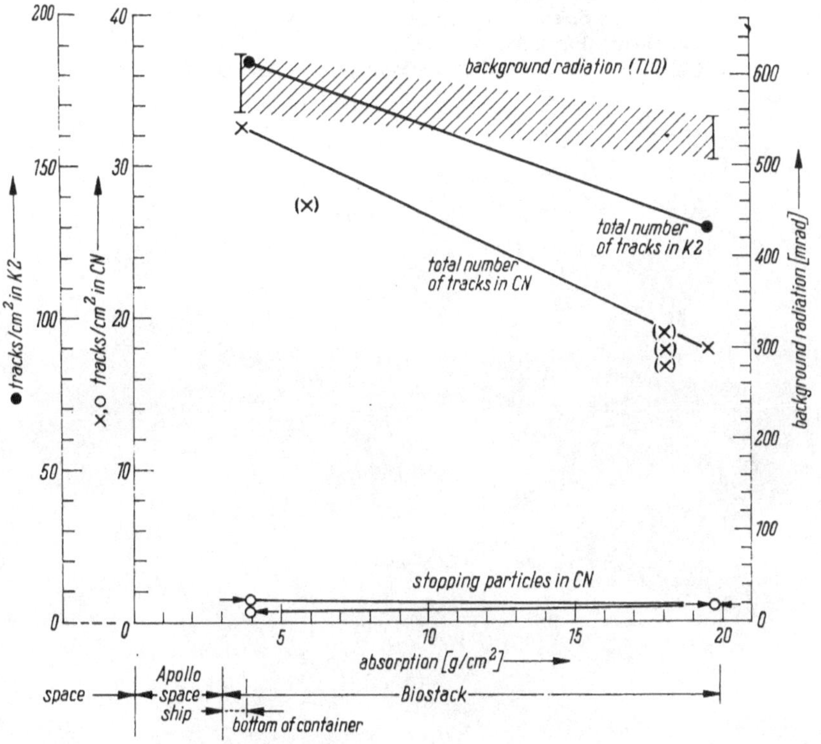

Fig. 3. Data obtained from physical detectors in the Biostack I, flown aboard Apollo 16, as a function of absorption.

The biological actions of HZE particles must be considered in all space missions, both to estimate the radiation hazard in manned space flight, and to determine additional effects occurring in other biological space flight experiments (especially in experiments on possible biological effects of weightlessness). Such experiments, which will be mainly long-term ones, must be conducted on objects in a most sensitive stage of development, e.g. cellular division, cell differentiation, embryonic development. The results of the Biostack experiments are described in the papers that follow.

References

[1] H. B. Chase, J. Aviat. Med. **25**, 388 (1954).
[2] J. Eugster, Weltraumstrahlung. Medizinischer Verlag Hans Huber, Bern, Switzerland 1955.
[3] T. Brustad, Rad. Res. **14**, 139 (1961).
[4] H. Yagoda et al., Military Med. **128**, 655 (1963).
[5] W. Haymaker et al., Aerospace Med. **41**, 989 (1970).
[6] C. A. Tobias, J. Aviat. Med. **23**, 345 (1952).
[7] C. A. Tobias et al., Life Sciences and Space Research XI, 233 (1973).
[8] H. J. Schaefer, Health Phys. **23**, 327 (1967).
[9] H. Bücker et al., Life Sciences and Space Research XI, 295 (1973).

Life Sciences and Space Research XII — Akademie-Verlag, Berlin 1974

THE CHARGE SPECTRUM OF HEAVY COSMIC RAY NUCLEI MEASURED IN THE BIOSTACK EXPERIMENT ABOARD APOLLO 16 USING PLASTIC DETECTORS

W. Enge, R. Beaujean, K.-P. Bartholomä and K. Fukui[a]

Institut für Reine und Angewandte Kernphysik, Kiel, FRG

In the Biostack experiment flown aboard Apollo 16 plastic detectors were used to select those individual biological objects which were hit by single heavy cosmic-ray nuclei during the 266 hours of space flight. The energy loss deposited in the biological object can be determined. This method needs a separate calibration in every experiment, i.e. an individual track etching rate as function of the energy loss. This calibration is given.

An area of 36 cm^2 of eight cellulose nitrate sheets was scanned for tracks of heavy cosmic-ray particles that stopped in the sheets. About 200 particles could be evaluated. Calibration was achieved using the oxygen isotope ^{16}O of the cosmic-ray ions and the charge spectrum of all measured particles ($Z = 5-16$) could be obtained.

1. Introduction

A biophysical experiment (Biostack) was exposed aboard Apollo 16, launched on 16 April 1972, to heavy cosmic-ray nuclei in order to study their effects on living biological objects. The Biostack represents collaboration between scientists at biological and physical institutes in Europe under the principal investigator, H. Bücker, Arbeitsgruppe Biophysikalische Raumforschung, Frankfurt, FRG. It is a cylinder of 10 cm diameter and 10 cm height, consisting of more than 200 biological and physical layers, subdivided into 9 units (4 biological and 5 physical units).

A new method using plastic detectors has been developed to correlate single heavy cosmic-ray nuclei with single biological objects [1]. A short description of the method is given below. Calibration of energy loss determination is obtained using measurements on heavy cosmic-ray particles.

2. Correlation between Biological Objects that have been hit and Heavy Cosmic-ray Particles

The biological objects are imbedded in polyvinyl alcohol foil (PVA) which is cast on one surface of cellulose nitrate sheets (see Fig. 1, sheets 2 and 3). The biological objects are in contact with the physical detector during exposure to cosmic rays, during etching and during the analysis of the tracks under the micro-

[a] On leave from AFCRL, Bedford, Mass., USA.

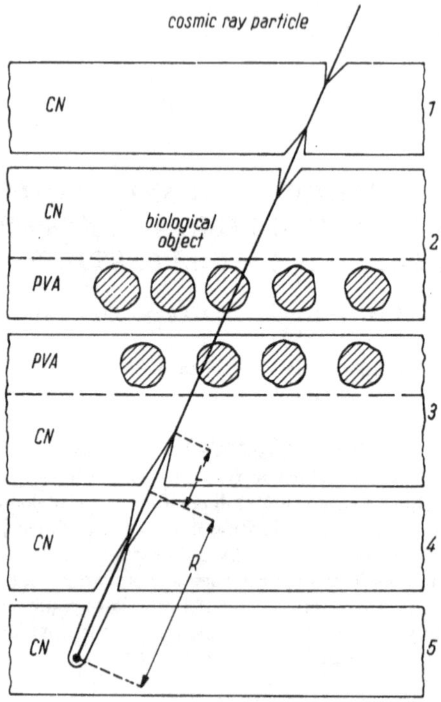

Fig. 1. A charged particle penetrating several sheets of the Biostack (after etching). CN cellulose nitrate sheet; PVA, polyvinyl alcohol.

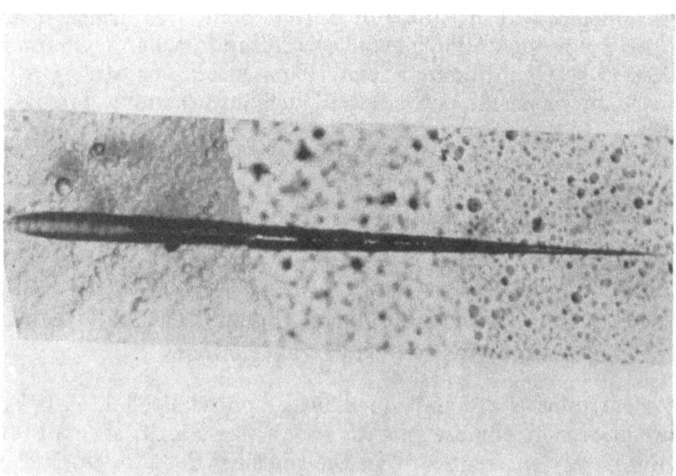

Fig. 2. Microphoto of an etch cone of a heavy cosmic particle after 15 hours etching. About 900 μm long.

scope. A special etching procedure for one side of a sheet only protects the biological side against the NaOH solution.

In Fig. 1 the trajectory of a cosmic-ray particle that has penetrated several sheets of the Biostack is shown (after etching). Fig. 2 shows a single etch cone of a heavy cosmic-ray particle. The plastic detectors in this experiment yield two kinds of information: (a) the localization and the direction of an etch cone indicates whether a biological object was hit or not and (b) the length of an etch cone gives the energy loss of the heavy particle at this specific part of its trajectory [2, 3]. The calibration of this method is described in § 4.1; for this a representative set of heavy cosmic-ray particles is necessary.

3. Measurements on Tracks of Heavy Cosmic-ray Particles

3.1. Experimental Procedure

In the physical unit A1 (which was closest to the outside, under shielding of about 3.5 g cm^{-2}) eight cellulose nitrate plastic detector sheets were analysed. 250 μm thick Daicel cellulose nitrate sheets of 10 cm diameter were etched in a mechanically stirred, ultrasonically agitated (6 ± 0.1) N-NaOH solution with

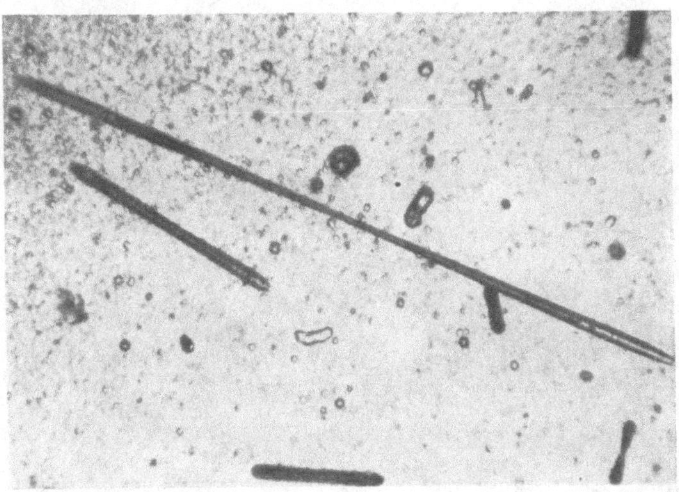

Fig. 3. Microphoto of several double cones etched through after 15 hours etching. The longest track is 2000 μm long, and that at the bottom of the figure 360 μm.

0.05% Benax surfactant at (40 ± 0.1) °C. Seven sheets were etched for 4 hours, one sheet for 15 hours. A new scanning method was used [4], in which two adjacent sheets were always scanned together in distilled bubble-free water under a stereo microscope. Fig. 3 shows several tracks which have been etched for 15 hours.

3.2. Evaluation of the Measurements

In an area of 36 cm^2 312 tracks were found which stopped in one of the five sheets. Among these are two categories of tracks which could not be evaluated: (i) all those tracks which had produced etched cylinders instead of two separate cones; (ii) all those tracks where the cone at the stopping end was still pointed, i.e. the cone had not been etched up to the very end of the trajectory of the particle. Corrections for these effects and for the visual scanning loss and registration loss have not yet been made.

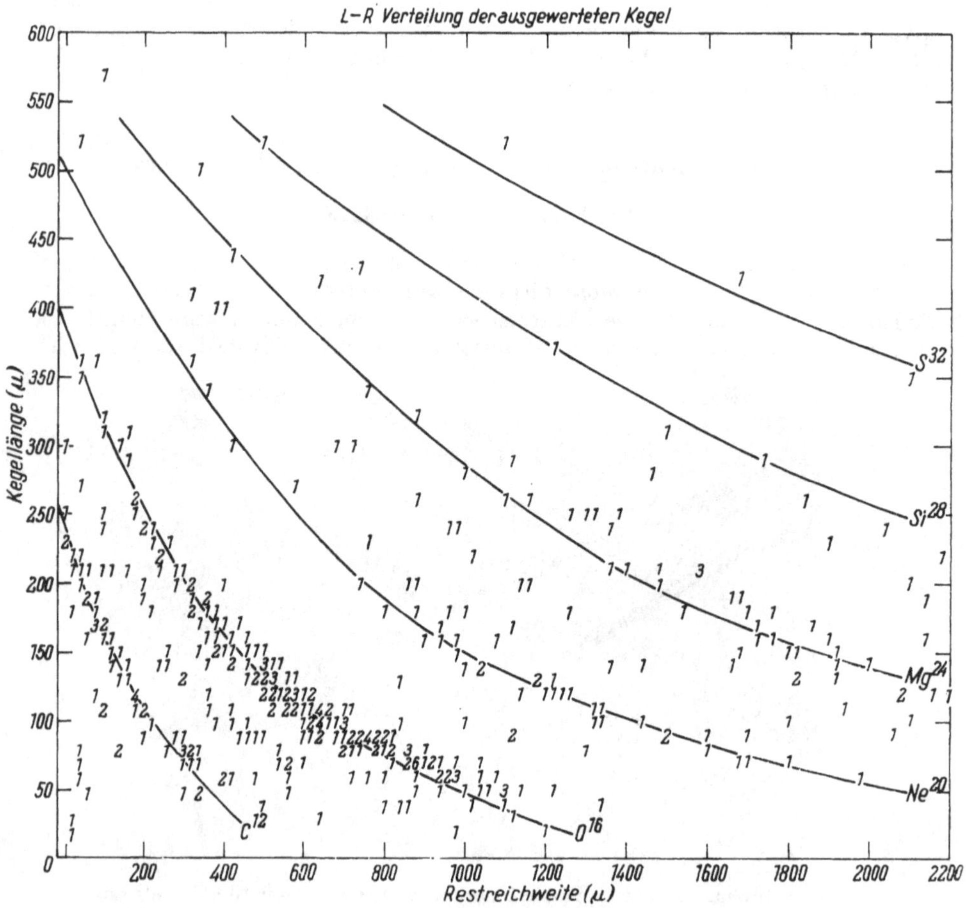

Fig. 4. The cone length is plotted against its residual range. The curves are calculated assuming all cosmic oxygen is the isotope ^{16}O.

All etch cones along a particle's trajectory were measured (1250× magnification) and the residual range was calculated, as described in more detail in [2, 3]. Thus 193 tracks could be analysed. For each particle, having one etch cone or more, each cone length L was plotted against its residual range R in an (L, R) diagram which is a $(dE/dx, E)$ plot (see Fig. 4). The different curves relate to different elements.

4. Results

4.1. Calibration

The etching behaviour of heavy ions from accelerators in Daicel cellulose nitrate has been studied [2]. This calibration can be transferred to measurements on cosmic-ray ions, but only with some changes because of the different exposure conditions. Here calibration using cosmic-ray oxygen ions was made, assuming that all oxygen is the isotope ^{16}O. The amount of air or free oxygen during the exposure seems to play a role that is not negligible, resulting in different track etching rates. These differences have been noticed already earlier [2] between accelerator exposures (5—10 minutes in vacuum) and balloon exposures (10 hours in 1—5 g cm^{-2} air). But with the present experiment, using exposure in a closed container (270—300 hours in air with a higher percentage of oxygen and some helium contamination) the effect has shown up much more obviously. The track etching rate which was measured in this work is the most sensitive one we have ever found in Daicel cellulose nitrate. This rate differs from those found in accelerator or balloon measurements by an amount up to 50%. For all biological evaluations this individual calibration, which is specific for this experiment, is essential.

4.2. Charge Spectrum

Using this individual calibration together with the theory of restricted energy loss (REL) criterion all other curves in Fig. 4 have been calculated, as described in [2]. These curves fit very well to the maxima of the strips and to the direction of the cones of the same particle (see Fig. 4, e.g. curve S^{32}). Each track in the (L, R) plot, having one cone or more, has been allocated to the element with the nearest curve, and in this way the charge determination shown in Table 1 was

Table 1

Measured Abundances of the Individual Elements

Element	Li_3	Be_4	B_5	C_6	N_7	O_8	F_9
Abundance	2	4	5	37	16	71	1
Element	Ne_{10}	Na_{11}	Mg_{12}	Al_{13}	Si_{14}	P_{15}	S_{16}
Abundance	15	7	24	2	5	—	4

obtained. As mentioned above the abundances of the individual elements have not yet been corrected for any scanning effect, but it can be seen already that there are no big changes in the element abundances compared with measurements at balloon altitudes.

Acknowledgments

The authors are grateful to Prof. Dr. E. Bagge for the facilities of the Institute. They want to thank the other scientists of the Biostack experiment who together made this collaboration possible. This work was supported in part by the Deutsche Forschungsgemeinschaft.

References

[1] W. ENGE, K.-P. BARTHOLOMÄ, R. BEAUJEAN, W. HEINRICH, K. FUKUI, G. HORNECK and H. BÜCKER, Proc. 8th Int. Conf. Nucl. Photography and Solid State Track Detectors, Bucarest 1972, Vol. I, p. 558.
[2] R. BEAUJEAN and W. ENGE, Z. Phys. **225**, 416 (1972).
[3] K.-P. BARTHOLOMÄ, W. ENGE and K. FUKUI, Z. Phys. **259**, 75 (1973).
[4] K. FUKUI, Proc. 8th Int. Conf. Nucl. Photography and Solid State Track Detectors, Bucarest 1972, Vol. II, p. 386.

Life Sciences and Space Research XII — Akademie-Verlag, Berlin 1974

EXPERIMENTAL METHODS OF CORRELATION BETWEEN THE TRAJECTORIES OF COSMIC HEAVY IONS AND BIOLOGICAL OBJECTS: DOSIMETRIC RESULTS FROM THE BIOSTACK EXPERIMENT ON APOLLO 16 AND 17

R. Pfohl, R. Kaiser, J. P. Massué and P. Cüer

Centre de Recherches Nucléaires, Laboratoire de Physique Corpusculaire, Strasbourg-Cronenbourg, France

The biological objects in the Biostack are stacked alternately between physical detectors of HZE particle tracks which include nuclear emulsions. The pattern of the biological objects is transferred to the upper side of the adjacent emulsion by optical illumination. On each sheet of nuclear emulsion a coordinate grid is transferred to the bottom side by optical illumination. The visible track left by the passage of the heavy ion can then be very clearly localized. The charge of the ions is determined by photometric measurements of the tracks. These measurements are calibrated with known tracks of heavy ions accelerated at the Bevatron at Berkeley. The results are given as: flux, number of stars per cm^3, charge of the ions, comparison with other detectors (plastics, AgCl crystals, LiF), energy loss, hit region in the biological objects.

1. Introduction

The part of the Laboratoire de Physique Corpusculaire of the University of Strasbourg in the Biostack experiments aboard Apollo 16 and 17 is concerned with the detection and localization of cosmic heavy ions in order to identify which biological objects have been affected by these ions.

2. Experimental Methods

The biological objects are stacked alternately between physical detectors of HZE particles tracks which include nuclear emulsions (Ilford K2 and K5). The Biostack was disassembled in Strasbourg and during disassembly the pattern of the biological objects positioned directly over a K2 nuclear emulsion was transferred to the upper side of the emulsion by optical illumination. This illumination took place when the biological layer was still stacked on the Biostack support bolts and pressed into its supporting emulsion sheet. On each sheet of nuclear emulsion a coordinate grid was placed on the bottom side by optical illumination (Fig. 1). After disassembly, each biological layer that was stacked over a K2 nuclear emulsion was photographed to identify the exact position of the biological objects with respect to the emulsion. The visible track left by the passage of the heavy ion is thus very clearly localized. The development of the nuclear emulsions was made by a special technique for thick emulsions, the temperature process.

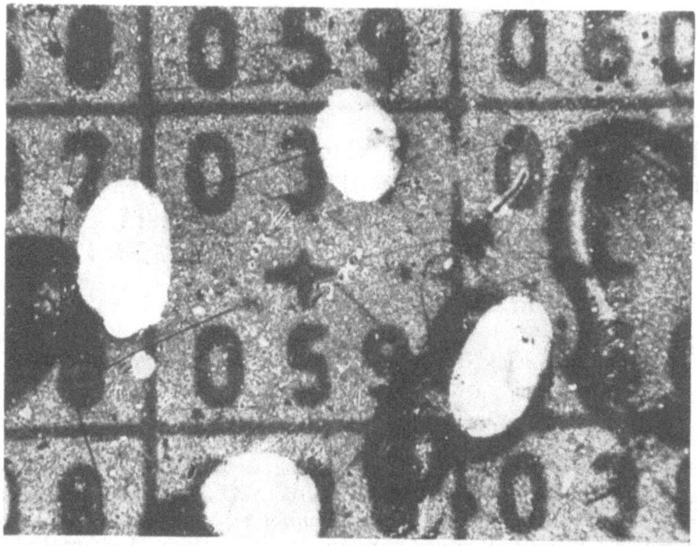

Fig. 1. Emulsion with coordinate grid.

The following observations were then made: tracks of the heavy ions whose positions were identified on the grid, together with measurement of their length and angle to the vertical (Fig. 2). Elements for the photometry were also calculated, e.g. length of a projected cell of 50 μm (to determine shrinkage); a Leitz MPV photometer was used for photometry of the ions. The thinning down of the track of heavy ions at the end of their path is a characteristic parameter for determining their change (Fig. 3) [1]; differentiation between ions requires an accurate knowledge of the known ions. The Bevatron at Berkeley produced beams of ^{12}C and ^{16}O of 250 MeV/amu and 2.1 GeV/amu, and these beams enabled us to make exact calibrations for the photometry of heavy ions (Fig. 4). Another problem is to follow tracks through the Biostack. A few thousand micrometres

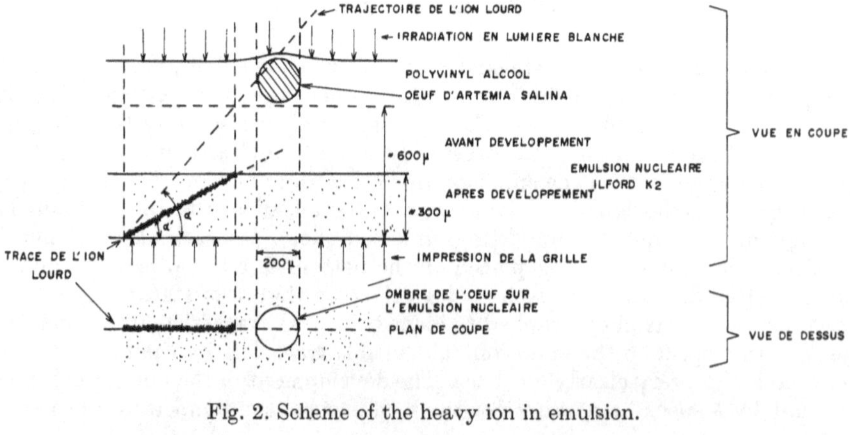

Fig. 2. Scheme of the heavy ion in emulsion.

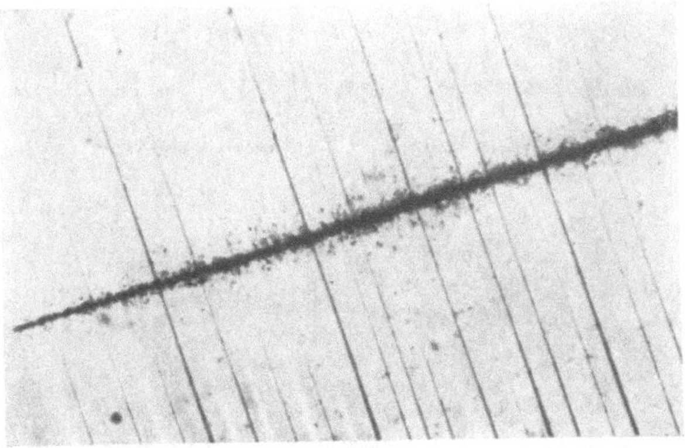

Fig. 3. Heavy ion ending in the emulsion.

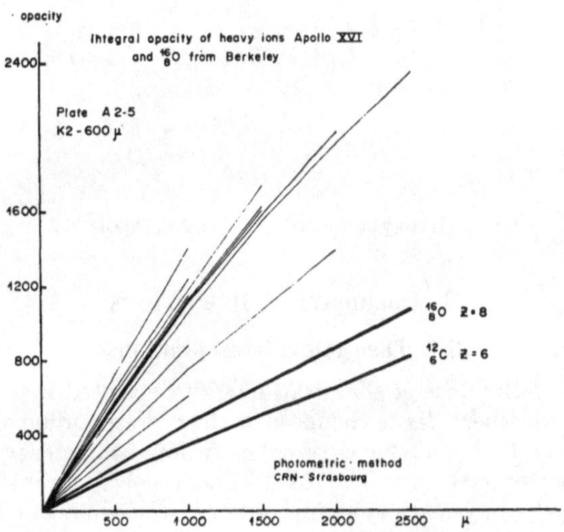

Fig. 4. Integral opacity of heavy ions, Apollo 16. ^{12}C and ^{16}O from Berkeley.

separate two emulsions, but by our method we were able to follow tracks for 66000 μm. It is important for dosimetry to know if a track is near the end or not. Fig. 5 is a picture from a track in Apollo 17 in the last 3000 μm of its range.

The areas scanned and observed fluxes were:

Apollo 16: area scanned 26321 mm² (8 plates)

No. of tracks ending 558	Flux 2.1 cm^{-2} (35 cm^{-3})
(H + M) 13744	Flux 52 cm^{-2}

Apollo 17: area scanned 11047 mm² (6 plates)

No. of tracks ending 297	Flux 2.7 cm^{-2} (45 cm^{-3})
(H + M + L) 12449	Flux 113 cm^{-2}

H + M: heavy ions of charge $\geqq 6$; H + M + L: heavy ions of charge $\geqq 3$

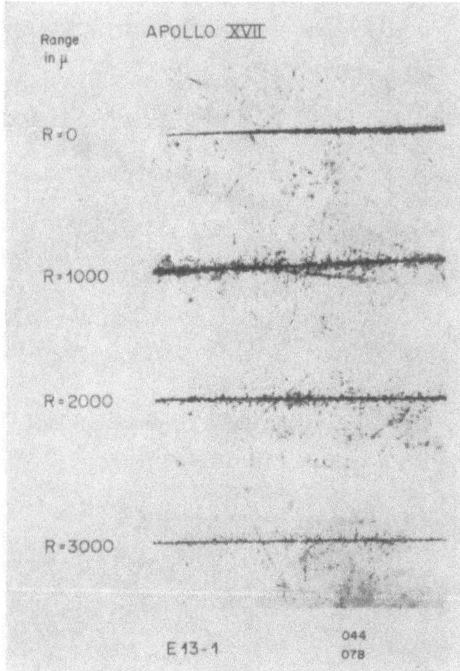

Fig. 5. Heavy ion at several ranges, Apollo 17.

3. Dosimetry of Heavy Ions

3.1. Theoretical Considerations

The definition of the dose is the mean energy deposited in a given volume of matter: $D = dE/dm$ where dm is the mass of the corresponding volume. The unit is the rad (1 rad $= 10^{-2}$ joule kg^{-1}). For heavy ions the energy deposited along the track is a better measure. The LET (linear energy transfer) can be used corresponding to the physical "stopping power" dE/dx measured in MeV cm^{-1} or MeV cm^2 g^{-1}:

$$1 \text{ MeV cm}^{-1} = \varrho \text{ MeV cm}^2 \text{ g}^{-1}$$

where ϱ is the density. It is useful to introduce the REL (restricted energy loss) REL $= (\text{LET})_\Delta$, in which the energies transferred by δ rays up to a given energy Δ are considered. The basis of theoretical calculations is the well-known Bethe-Bloch formula. At any given velocity (or at the same energy/nucleon) we have the following relations between protons and heavy ions:

$$E_M(\beta) = M \cdot E_p(\beta) \text{ energy}$$

$$R_M(\beta) = \frac{M}{Z^2} R_p(\beta) \text{ Range}$$

$$\left(\frac{dE}{dx}\right)_M (\beta) = Z^2 \left(\frac{dE}{dx}\right)_p (\beta) \text{ LET}$$

where $E_p(\beta)$ is energy of the proton at velocity $\beta(= v/c)$ and $R_p(\beta)$ is the residual range of the proton in emulsion at velocity β.

At the end or their range the LET curve is of Bragg shape. Fig. 6 is the curve for ^{16}O in emulsion. The radial structure of the heavy ion tracks has been investigated in many laboratories. The formula is given by

$$\left(\frac{dE}{dr}\right) = \frac{Ne^4}{mc^2 W} \frac{Z^{*2}}{\beta^2} \left(\frac{1}{W} - \frac{1}{W_{\max}}\right)$$

(W = energy of δ rays, Z^* is the effective charge)

The radial dose is proportional to Z^2 and $1/W^2$.

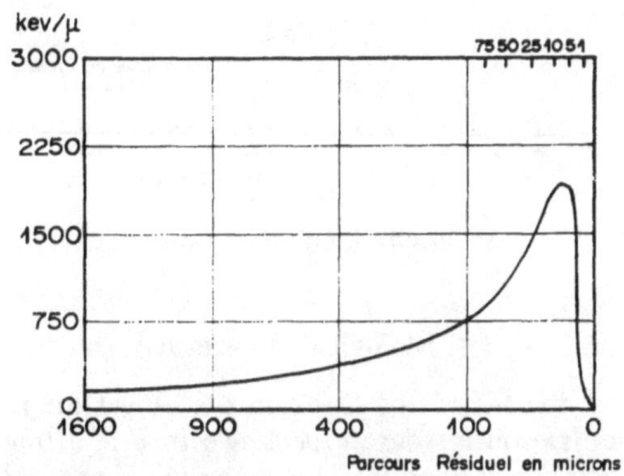

Fig. 6. Bragg curve for ^{16}O.

3.2. Calibration

The calibration was made against the ions of the Berkeley Bevatron. Fig. 7 indicates the opacity against the range for ^{12}C and ^{16}O ions. The correlation between opacity and ionization is very important for dosimetry (Fig. 8). For very high energy all the experimental points lies on the same curve.

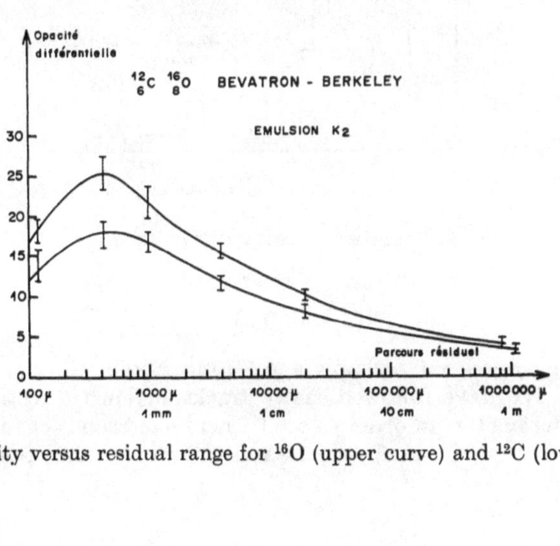

Fig. 7. Opacity versus residual range for ^{16}O (upper curve) and ^{12}C (lower curve).

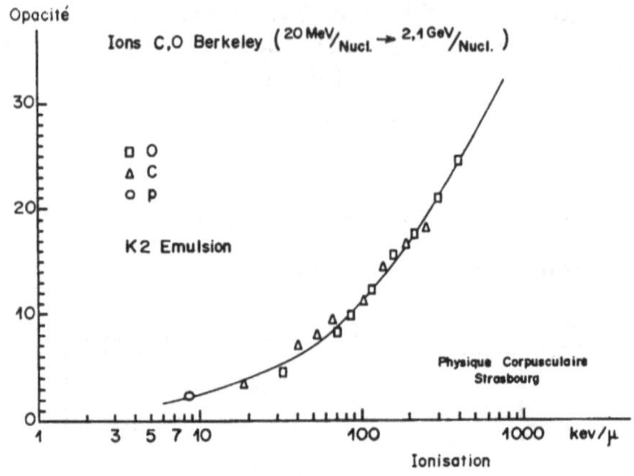

Fig. 8. Opacity—ionization calibration.

3.3. Biological Objects Hit

We also calculated the length of track in the object hit in sample A4-8 (*Artemia salina*). The ionization histogram is shown in Fig. 9 (measured from emulsion A4-7). The mean energy deposited for an egg is of the order of 50 MeV.

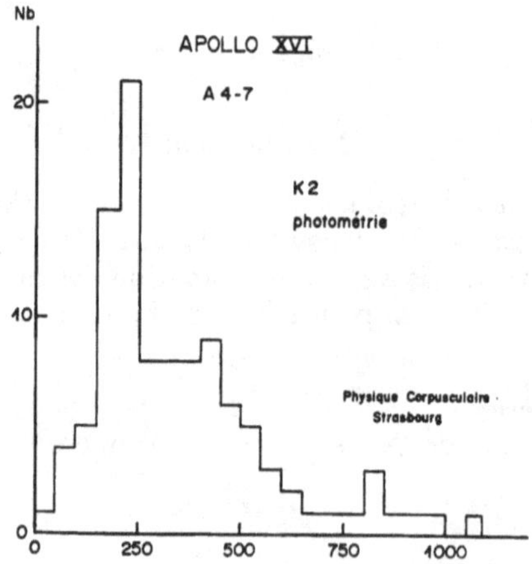

Fig. 9. Ionization of heavy ions in plate A4-7.

3.4. Dose

For calculating doses we tock plate A5-15 in Apollo 16. Fig. 10 is the histogram obtained. We have followed many tracks in both directions over many thousands of micrometres in order to see if there has been a change in ionization or not. The dose observed in Apollo 16 was 3.3 mrad; since the total dose was

650 mrad, this is equivalent to 0.50%. For Apollo 17 in plate E 7-1 the result was 3.5 mrad, i.e. 0.48% of the total dose of 730 mrad. As we have seen theoretically the action is restricted to a short cylinder along the path of the ion. If we assume that the total energy is deposited in a cylinder of 10 μm radius the local dose is not 3.3 mrad but 20 rads.

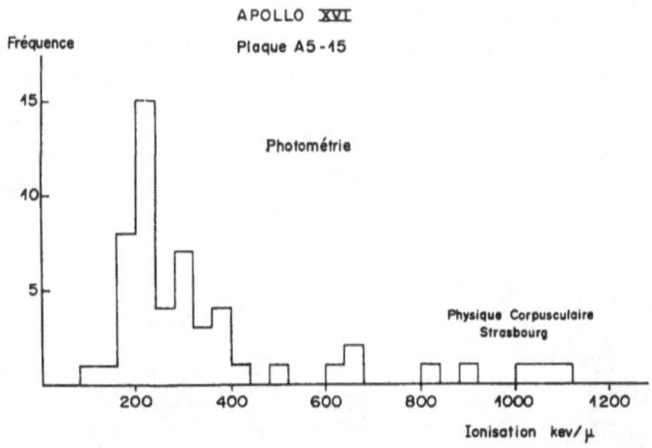

Fig. 10. Ionization of heavy ions in plate A5-15 of Apollo 16.

3.5. Nuclear Stars in Emulsions

The background of cosmic rays interacts with the nuclear emulsion and gives many stars. These stars give rise to an energy deposition which is by no means negligible. There are up to 15000 stars per cm³ in K5 emulsion in Apollo 16. A short survey indicated a mean of 7 branches per star (4 black tracks, 2.5 grey, 0.5 white) which deposit about 42 MeV per cm³. The dose is 2.6 mrad. If one calculates the local dose by assuming this is deposited in a sphere of 10 μm diameter the local dose is 3 rad, about 10^3 times higher than the mean dose. For comparison we give balloon flight experiment results from 1970.

20400 m	1850 stars · cm^{-3} day^{-1}	G5
26000 m	1340 ,,	G5
34000 m	900 ,,	G5
Apollo 17	950 ,,	K2

These doses are similar to those in other space experiments.

Life Sciences and Space Research XII — Akademie-Verlag, Berlin 1974

AgCl DETECTORS IN THE BIOSTACK II EXPERIMENT ABOARD APOLLO 17

G. Henig[a], E. Schopper[a], J. U. Schott[a] and W. Rüther[b]

[a] Institut für Kernphysik, University of Frankfurt/Main, FRG
[b] Klinik und Poliklinik für Nuklearmedizin, University of Marburg, FRG

Two layers of AgCl detectors with a total surface of 90 cm² were flown. Tracks of nuclei, from light ($Z > 4$) up to the heaviest were recorded and could be distinguished by their geometrical trackwidths. The tracks were divided into five groups of atomic numbers, and their abundance was measured. Also the number of (surviving) nuclear stars was counted. 22.5 cm² of the detector surface were covered with eggs of *Artemia salina*. The detectors could be developed without removing the eggs, so that the spots hit could be determined directly. The radiation effect on these eggs is being investigated.

In the Biostack II experiment 24 single AgCl detectors with a total surface of 90 cm² and a volume of 1.8 cm³ were flown for 305 hours in complete darkness. 240 hours later, on 29 December 1972, they were exposed to yellow light in order to stabilize the latent tracks of recorded nuclear particles. The AgCl detectors normally require irradiation with yellow light during the passage of particles for the stabilization of their tracks [1]. The quick fading of latent tracks, if they are not stabilized, allows one to select the time during which tracks are recorded. This facility for illumination was not available on Apollo 17, so we had to use detectors in which the tracks faded much less.

A study of the same type of AgCl crystals irradiated with 6.1 MeV α-particles before the flight showed that after storage times of 240 and 545 hours, between the flight and their development, these α-tracks faded in the following manner: The total blackness, as integrated across the density profile of the track, de-

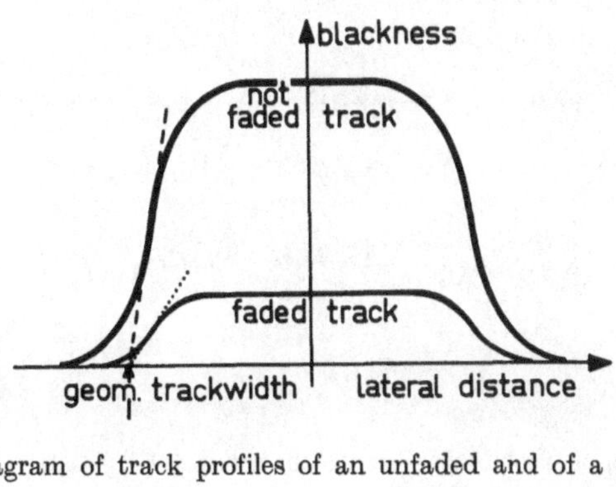

Fig. 1. Schematic diagram of track profiles of an unfaded and of a partially faded track.

creased to values of about 25% and 15% respectively, while the geometrical width of the track remained almost constant (see Fig. 1). Although the track density therefore did not yield a good estimate, the geometrical track width does allow

Table 1

Group	geometrical trackwidth (μm)	estimated charge Z	particle number/cm²
L (light particles)	< 0.8	$\leqq 6$	254 ± 14
M (medium)	0.8–1.2	6– 9	68.4 ± 6.4
H (heavy)	1.2–2.0	9–15	17.3 ± 1.9
V (very heavy)	2.0–3.0	14–26	4.4 ± 0.8
S (very very heavy)	> 3.0	> 24	2.4 ± 0.7

a rough estimate of the charge of the particles. On account of the threshold of sensitivity at about ten times the minimum ionization for protons, tracks of particles with charge $Z \leqq 4$ are not recorded in the region of relativistic energies. The surviving tracks of particles were divided into five groups of atomic numbers, and their abundance was measured (Table 1). Fig. 2 shows an example for each of these groups of tracks.

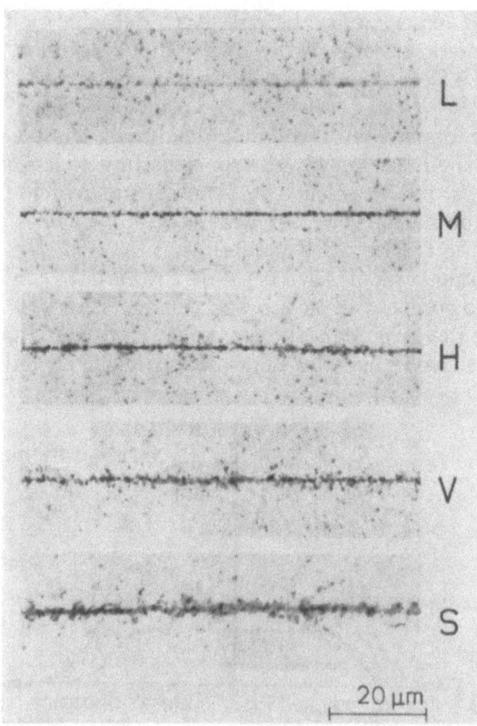

Fig. 2. Examples of tracks belonging to the groups of L, M, H, V and S particles.

The number of surviving nuclear stars was found to be 4500 per cm³ of AgCl, estimated from 471 stars observed. In default of adequate stabilization of the AgCl detectors some fading of particle tracks, in particular of protons from stars, had to be accepted.

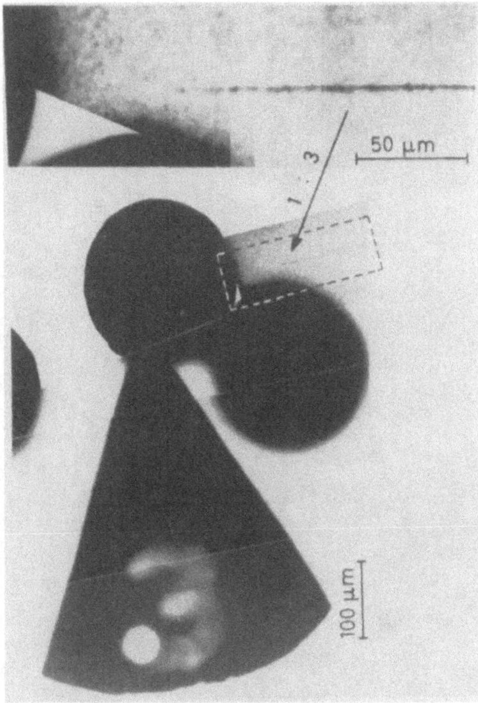

Fig. 3. Two eggs of *Artemia salina* (large black circles), one of them (left-hand circle) hit by a heavy particle and marked by a numbered triangle (3); the track is shown in dotted triangle. Above is a 3:1 magnification of the track and of the edges of the eggs.

Six of the crystals were covered with eggs of *Artemia salina*. These crystals could be developed with blue light (410 nm) without removing the eggs. 79 eggs that had been hit were found, most of them hit by particles of at least medium charge; these eggs were marked by printing a small numbered triangle on the surface of the crystal with white light by way of the side eyepiece of the microscope. The tracks of the particles were classified in the manner described. Fig. 3 shows such a marked egg, hit by a heavy particle. The marked eggs have been isolated and are being investigated after hatching.

Acknowledgments

The work was supported by the German Federal Ministry of Research and Technology (St. Sch. 504).

References

[1] E. Schopper, G. Haase, F. Granzer, K. Dardat, G. Henig, J. U. Schott, H. Bücker and G. Horneck, Life Sciences and Space Research XI, 215 (1973).

Life Science and Spaces Research XII — Akademie-Verlag, Berlin 1974

PRELIMINARY RESULTS ON THE ACTION OF COSMIC HEAVY IONS ON THE DEVELOPMENT OF EGGS OF *Artemia salina*

W. RÜTHER[a], E. H. GRAUL[a], W. HEINRICH[b], O. C. ALLKOFER[b], R. KAISER[c] and P. CÜER[c]

[a] Klinik und Poliklinik für Nuklearmedizin, Universität Marburg, Marburg/Lahn, FRG
[b] Institut für Reine und Angewandte Kernphysik, Universität Kiel, Kiel, FRG
[c] Centre de Recherche Nucleaire, Strasbourg-Cronenbourg, France

The Biostack experiment (Principal Investigators: H. Bücker) aboard Apollo 16 contained one unit with eggs of the brine shrimp *Artemia salina*. The eggs were fixed in polyvinyl alcohol foils. Nuclear emulsions and plastics were used as track detectors. The development of 260 eggs hit by cosmic heavy ions was investigated. It differed significantly from the development of the flight controls—eggs flown in the Biostack but not hit by heavy ions—and of the ground controls. From these results it was concluded that a hit of a single heavy ion may injure the encysted blastula. This damage was found to influence the gastrula formation and even the hatching process of the nauplius. Abnormalities of the orthonauplius were observed during the development of the hit eggs. These are shortened extremities, or abnormal thorax or abdomen.

In addition, in the Biostack II aboard Apollo 17 eggs of *Tribolium confusum* and *Carausius morosus* were included. The influence of single heavy ions on the development process of these highly organized insects was investigated.

1. Evaluation of the Track Detectors

The unit contains four biological layers coated with cellulose-nitrate and four biological layers coated with nuclear emulsion. In this report attention is given to the heavy ion tracks with high LET values. The etched cones in the plastic layer (A4-32) were measured for possible hits and the point where the track passes through the PVA layer was calculated. The energy loss of the heavy ion was determined from the length of the etched cones in that layer and from the cones in the next layer. Table 1 shows the LET-distribution of hits in layer A4-32.

Table 1
LET Distribution of Hits in Layer (A4-32)

LET (10^3 MeV g^{-1} cm^{-2})	Number of hits
$\geqq 1$	26
$\geqq 2$	14
$\geqq 2.5$	4

2. Evaluation of the Biological Layers: Biological Radiation Investigation on *Artemia salina*

The development of *Artemia salina* is dependent on the temperature and the salt concentration, but these two components of the environment may vary within wide limits. In the extreme limits different combinations of these components influence the development, growth, reproduction etc. The salt concentration of the medium has an influence on the hatching process, caused by the fact, that the outer osmotic pressure restricts the possibility of the hydration phase. The cyst can only swell if the salt concentration is less than 9%. At higher salt concentrations the shell remains semi-concave even if the eggs remain for a longer time in the salt water medium [1—5]. The requirements of food for *Artemia* can easily be satisfied by the alga *Dunaliella viridis*; therefore it is not difficult to culture single shrimps or even large numbers over a longer time.

In the first investigations (Apollo 16) the eggs of *Artemia salina* hit by heavy ions showed the following biological radiation effects. On the one hand the number of hatching nauplii was essentially lower (14.5%) than for the eggs of the ground controls (90%) and for the eggs flown in the Biostack but not hit by heavy ions (46.2%). On the other hand the nauplii that developed out of the eggs hit by heavy ions showed a high lethal rate in the first four days after hatching (Table 2 and Fig. 1).

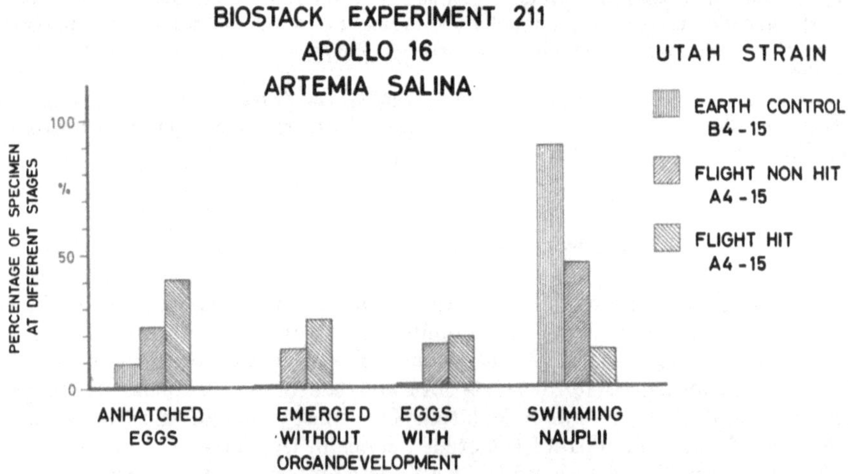

Fig. 1. Development percentages of *Artemia salina* eggs hit by heavy primaries; in flight but not hit; earth control.

The growth of the animals which survived the first days after hatching is shown in Fig. 2. The nauplii hatched out of the eggs of ground controls increase in length continuously, whereas the nauplii hatched out of eggs hit by heavy ions were stunted in growth (6 mm) and did not even reach the normal length of 12 mm 25 days after hatching. Also the nauplii which developed from eggs not hit by heavy ions, but flown in the Biostack, showed a retarded development until the 13th day and then grew to their normal length. Furthermore a relatively late maturity was observed, taking for this the pair mating as a criterion (Fig. 3).

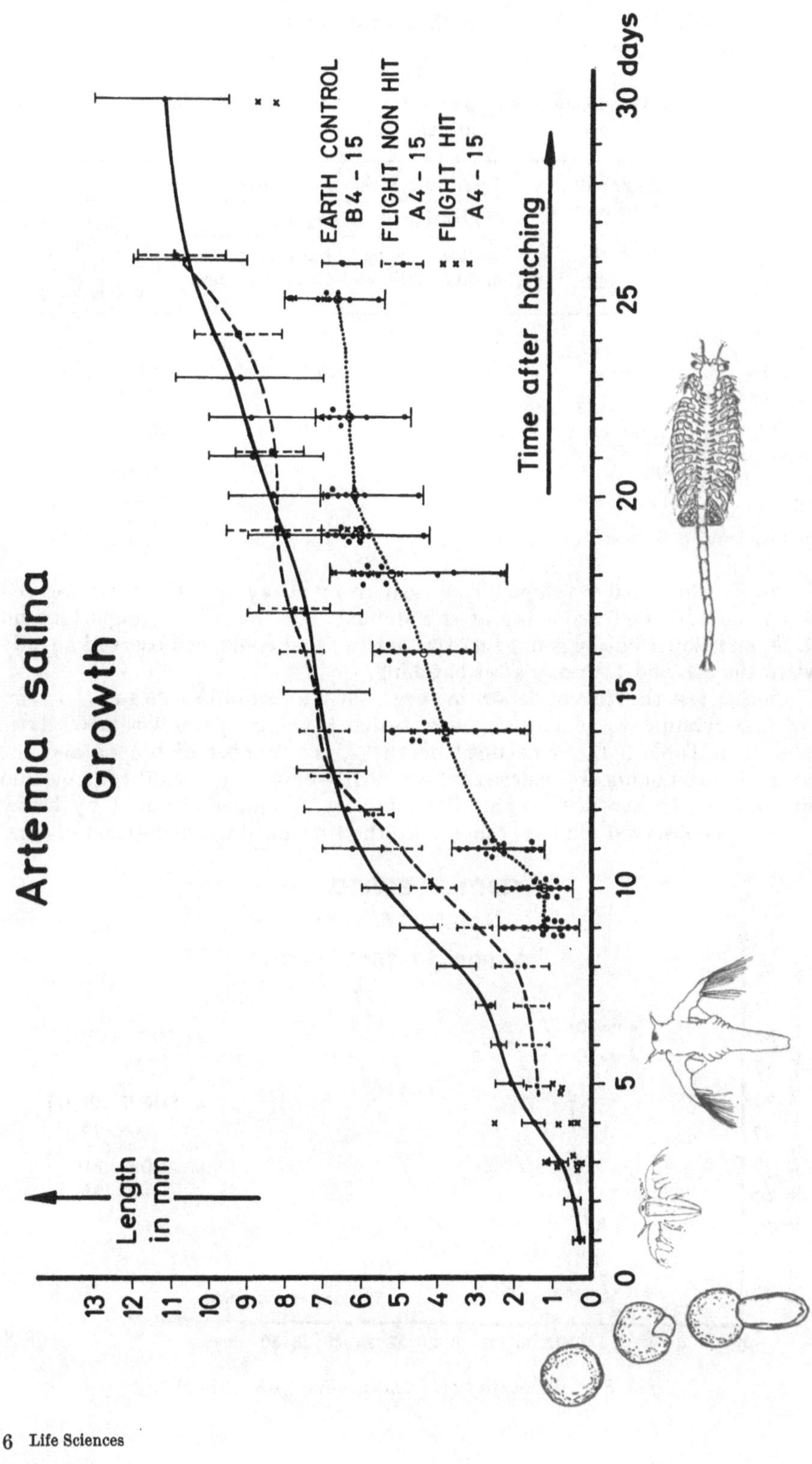

Fig. 2. Growth of *Artemia salina*.

Table 2

Frequency (%) of the Development Stages after Irradiation Damage by Heavy Primaries and at Controls

	Eggs without development	Eggs with development (egg shell cracked)		
		No tearing of the egg membrane		Tearing of the egg membrane Swimming nauplius larva
		No basic shape of larva formed	Basic shape of larva formed	
Group	1	2	3	4
Eggs hit by heavy primaries	40.9	25.5	19.1	14.5
Eggs not hit by heavy primaries, background irradiated	23.0	14.5	16.3	46.2
Ground controls	9.3	0	0.7	90.0

For *Artemia* which had developed from eggs hit by heavy ions, this was observed between the 12th and 22nd day after hatching. For the other groups (ground controls and flown but eggs not hit) the pair mating could be observed already between the 8th and 11th day after hatching.

In another test the time of depositing eggs, the number of broods and the number of descendants (F_1 generation) were tested for single pairs within 60 days. As shown in Table 3 the time until maturity, the number of broods and the number of descendants (F_1 generation) are nearly equal for ground controls and flight controls. In contrast to this the maturity of animals injured by heavy primaries was observed relatively late. Also the time until the deposition of eggs,

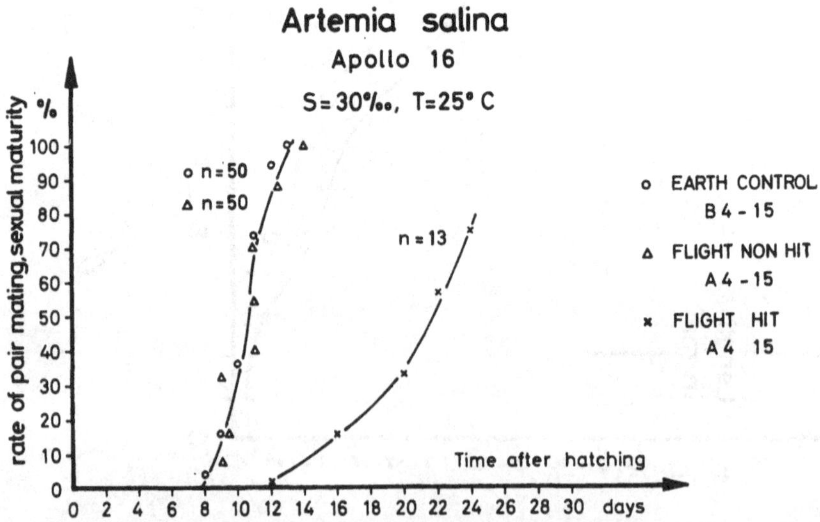

Fig. 3. Sexual maturity of *Artemia salina* (pair mating).

Table 3

Artemia salina. Development and Reproduction

	Sex. maturity, days after hatching (mating)	Days up to egg-production	No. of nauplii per brood and female in 2 months*	No. of nauplii in two months
Earth control	8–13	5– 9	(3) 132–<u>148</u>–160	359–<u>394</u>–423
Flown but not hit	9–14	5–10	(3) 105–<u>126</u>–148	225–<u>251</u>–287
Flown and hit	12–24	20–24	(0–2) 0–<u>82</u>–112	0–<u>102</u>–169

Salt concentration = 3%, T = 25 °C

* Number of broods in brackets.

Underlined figures are the arithmetic middle points.

when eggs were produced generally, is nearly double that for animals of ground control. The number of broods fluctuated from 0 to 2. Accordingly the number of descendants in the F_1 generation was essential lower (102).

Anomalies in the body and the extremities could be observed more frequently for the nauplii which had developed from eggs hit by heavy primaries (6%). Similar anomalities are relatively rare for ground controls and flight controls (0.5%).

The previous reports dealt with the development from eggs to larvae. It was found that only a small number of eggs, relative to the ground controls (90%), developed to swimming nauplius larvae, both for eggs hit by heavy ions (14.5%) and for eggs not hit by heavy ions but flown in the Biostack (46.2%). The further development of the larvae as reported here is different for these two groups. From the eggs which were not hit by heavy ions, but which were also exposed to the background radiation and to acceleration, vibration and weightlessness during the flight, animals developed which grew to the normal length of 12 mm, reached sexual maturity in nearly the same time as the ground controls and had a number of descendants comparable with those of the ground controls, whereas for the small number of animals which developed from eggs hit by heavy ions the further development, the time of maturity and the number of descendants differs significantly from the ground controls (Table 3).

To sum up it can be said that there was a significant difference in the development of the nauplius larvae from eggs of *Artemia salina* hit by heavy ions, from eggs with background irradiation and from eggs of the ground controls. From these results the radiation damage of the encysted blastula can be assessed [6–9]. This damage has in part an influence on the gastrula and even on the hatching process of the nauplius. But even many nauplii hit by heavy primaries do not survive the first casting of the skin with a high mitose rate. It is known from experiments [10–12] with heavy ions ($^{2+}$He, ^{6}C^{+}, ^{8}O^{+}, ^{17}A^{+}) that the dosage-effect curves are exponential. Assuming that for a sigmoidal dosage-effect curve some of the irradiated cells will retain their function, allowing survival of the organism, then an essential part of the cells, with their function of life may be destroyed by a single hit by a heavy particle, and have an adverse influence on embryogenesis. But one has to take into consideration that other factors (e.g. stress, such as vibration, acceleration etc.) can also have an adverse influence on development under irradiation by heavy ions or background irradiation. This is confirmed by the results of investigations of those eggs not damaged by heavy primaries.

References

[1] H. E. HINTON, Ann. Nat. Hist. **7**, 158 (1954).
[2] R. H. JENNINGS and D. M. WHITAKER, Biol. Bull. **80**, 194 (1941).
[3] —. BALLARDIN and P. METALLI, Current Topics in Radiat. Res. Quart. **7**, 181 (1972).
[4] Y. H. NAKANISHI, T. IWASAKI, T. OKIGAKI and H. KATO, Proc. Jap. Acad. **39**, 306 (1963).
[5] D. M. WHITAKER, J. Exptl. Zool. **83**, 391 (1940).
[6] S. T. BOWEN, Biol. Bull. **125**, 431 (1963).
[7] D. S. GROSCH and H. E. ERDMANN, Biol. Bull. **108**, 277 (1955).
[8] H. LANGHAM, Radiobiological Factors in Manned Space Flight, Nat. Acad. Sciences Nat. Res. Council, Washington, D.C., 1967.
[9] T. IWASAKI, J. Radiat. Res. **5**, 69 (1964).
[10] G. W. DOLPHIN and F. HUTCHINSON, Radiat. Res. **13**, 403 (1960).
[11] S. S. EASTER, JR, and F. HUTCHINSON, Radiat. Res. **15**, 333 (1961).
[12] F. HUTCHINSON and S. S. EASTER, Science **132**, 1311 (1960).

Life Sciences and Space Research XII — Akademie-Verlag, Berlin 1974

MICROBIAL STUDIES IN THE BIOSTACK EXPERIMENT OF THE APOLLO 16 MISSION: GERMINATION AND OUTGROWTH OF SINGLE *Bacillus subtilis* SPORES HIT BY COSMIC HZE PARTICLES

G. HORNECK[a], R. FACIUS[a], W. ENGE[b], R. BEAUJEAN[b] and K.-P. BARTHOLOMÄ[b]

[a] Arbeitsgruppe für biophysikalische Weltraumforschung, Universität Frankfurt, FRG
[b] Institut für Reine und Angewandte Kernphysik, Universität Kiel, FRG

Bacillus subtilis spores were flown in the Biostack experiment aboard the Apollo 16 command module. The spores embedded in plastic foils were stacked between physical track detectors. The energy loss spectrum of the heavy particles of cosmic radiation was determined. Biological studies were restricted to the high-energy loss component of these particles. Spores that had received single hits whose positions were determined with a typical accuracy of ± 1 μm, were investigated for radiation effects on germination and outgrowth. It was found that germination was not influenced by a hit by an HZE particle, but outgrowth was reduced significantly.

1. Introduction

The response of a single microbial cell to a passage of a very heavy particle of high-energy loss (HZE particle) of cosmic radiation was studied on spores of *Bacillus subtilis* 168 in the Biostack experiment (Principal Investigator: H. Bücker, University of Frankfurt) flown on the Apollo 16 command module. The biological layers, spores embedded in polyvinyl alcohol (PVA), were mounted on track detector sheets resulting in a fixed contact of the PVA layer with the cellulose nitrate detector sheet (CN). These sheets carrying the biological layer were stacked between pure plastic detector sheets (Fig. 1) in the typical arrangement of the biological units of the Biostack [1]. The *Bacillus subtilis* unit was accommodated in that region of the Biostack near the inner part of the command module. The measured fluence was about 17 particles cm^{-2} of $Z \geqq 4$ and a restricted energy loss (REL) > 0.8 GeV cm^2 g^{-1}.

2. Identification of Spores hit by an HZE Particle

Measurable tracks were obtained by etching the pure CN sheets with 6 n NaOH in an ultrasonic bath for 1.5 hours at 40 °C and the Lexan sheets for 8 hours at 70 °C. The CN sheets carrying the spore layers were etched only on the CN side at 30 °C without ultrasonic agitation [2] resulting in only one etch cone per track on the CN side (Fig. 1). For identification of the spore that had been hit the dip angle δ of the track of an HZE particle has to be known. Most of the microscopic measurements were done in the CN sheets adjoining the CN sheets bearing the spores. Microscopic measurements in the latter were avoided as far as possible

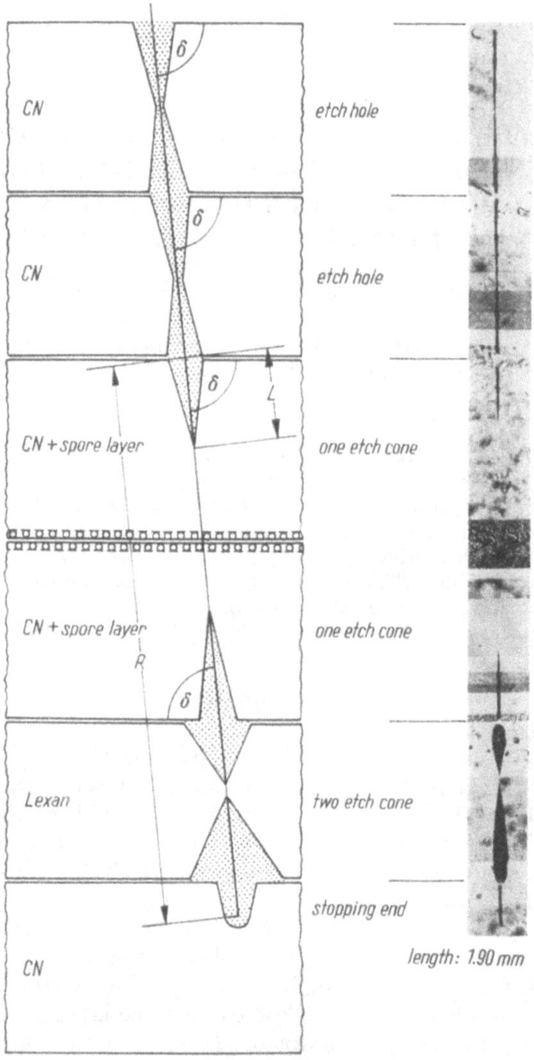

Fig. 1. Track of an HZE particle of cosmic radiation stopping in the *Bacillus subtilis* unit of the Biostack, microphoto (right) and sketch (left). δ, dip angle; L, length of the etch cone; R, residual range.

in order to prevent the spores from damage by illumination of long duration. Because of the small size (about 1.5 μm^3) of the spores the dip angle had to be measured with the highest possible accuracy. For this reason it was determined as a mean value of at least four independent measurements in the neighbouring CN sheets. The accuracy of the determination of the area hit in the biological layer is better than 2 μm^2. The length L of the single etch cone gives the REL of the particle at that point of its path (Fig. 1). The relation between this cone length and the residual range R determines the charge of the particle [3, 4]. Only particles with restricted energy loss (REL) higher than approximately 2 GeV cm^2 g^{-1}

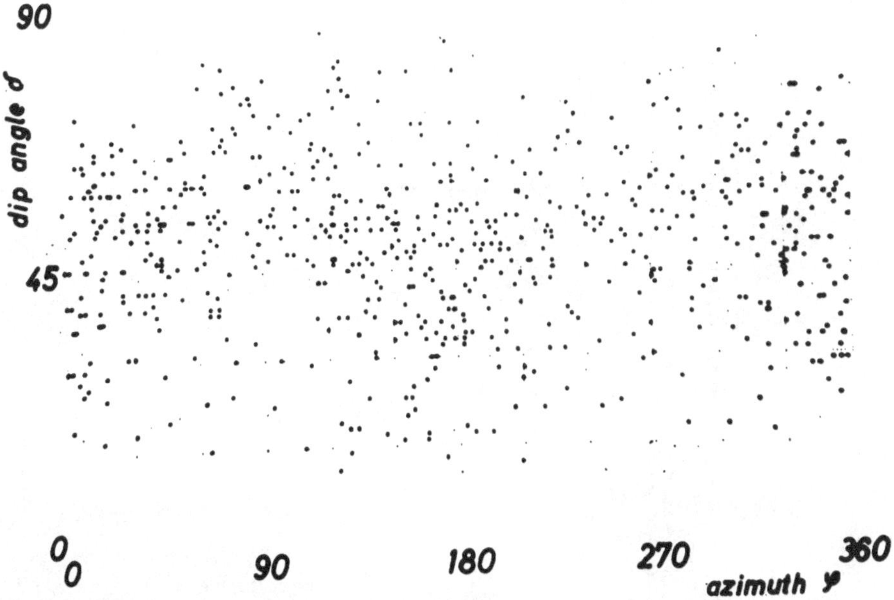

Fig. 2. Twofold angular distribution of heavy ions, measured in the *Bacillus subtilis* unit of the Biostack.

were selected for biological evaluation. A lower threshold of $Z = 12$ for their charge could be estimated from the restriction that REL > 2 GeV cm^2 g^{-1} for at least about 750 μm was required.

3. REL Spectrum of the Heavy Particles

A total of 670 tracks found in one CN sheet were evaluated to ascertain angle and energy loss distributions of the heavy particles. The twofold angular distribution for these particles obviously shows a deviation from isotropic distribution due to absorption by material surrounding the Biostack. Each point corresponds to one track (Fig. 2). For an isotropic distribution the frequency of points should vary symmetrically as sin 2δ with maximum density at 45°. Depending on the azimuth φ the tracks should be equally distributed. Thus no procedures to correct the measured fluences for scanning losses have been applied so far. The physical quantity of primary interest for biological studies is the REL of the heavy primaries. This was calculated from the length of the etch cones by use of calibration curves [5]. Fig. 3 shows the resulting integral distribution function. The experimental points between an REL of 0.8 and 2 GeV cm^2 g^{-1} are based upon measurements in CN. The cut-off at 2 GeV cm^2 g^{-1} is produced by the fact that, depending on the dip angle, above this value the etch cones from both sides of the detector surface started to overlap (see Fig. 1). In this case only a lower limit of REL could be assigned to the track. Experimental points above an REL of 3 GeV cm^2 g^{-1} correspond to measurements in an adjacent Lexan foil. Due to the lower sensitivity of this plastic detector these values range from 3 to 12.5 GeV cm^2 g^{-1} which is the highest REL found in Biostack I. The straight lines in Fig. 3 are adjusted sepa-

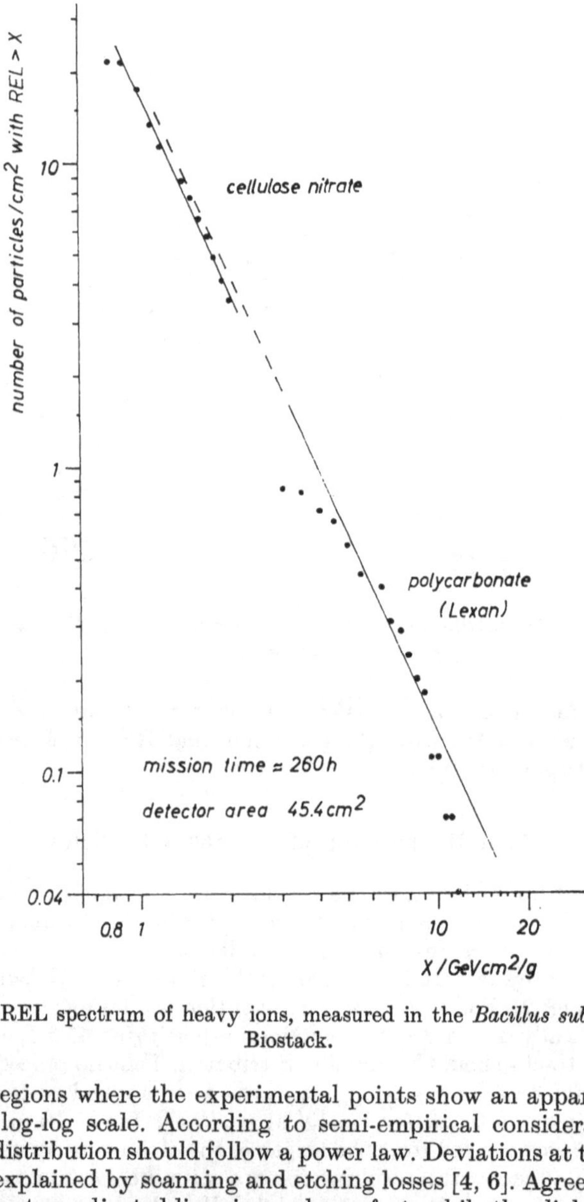

Fig. 3. Integral REL spectrum of heavy ions, measured in the *Bacillus subtilis* unit of the Biostack.

rately to the regions where the experimental points show an apparent linear behaviour on a log-log scale. According to semi-empirical considerations [6] the integral REL distribution should follow a power law. Deviations at the lower REL values can be explained by scanning and etching losses [4, 6]. Agreement between the slopes of the two adjusted lines is nearly perfect, while the slight difference in the absolute intensities may have its origin in the different modes used to obtain the calibration in CN and in Lexan. For CN an individual flight calibration was used [5], whereas for Lexan up to the present an accelerator calibration has been used. In Fig. 4 the REL distribution measured in the Biostack is compared with findings of Benton and Henke [7, 8] obtained from the microbial ecology evaluation device (MEED) experiment and the personal detectors of the astronauts of Apollo 16. Satisfactory agreement with our REL spectrum was found for the

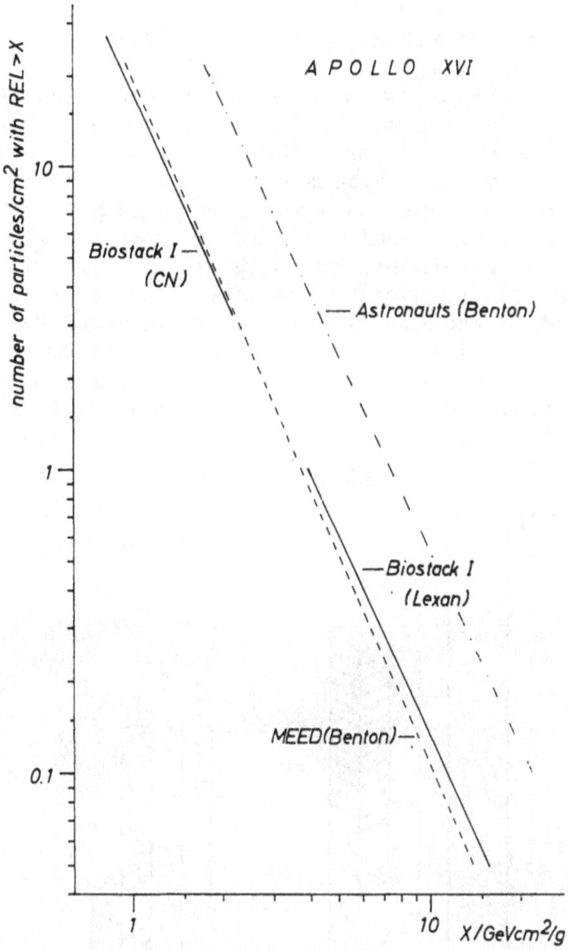

Fig. 4. Comparison of integral REL spectra obtained during the Apollo 16 mission.

MEED experiment which, like the Biostack, was placed within the command module for nearly the whole mission. The astronauts however had been exposed to higher fluxes when leaving the command module. Table 1 gives the numerical values for the different REL distributions shown in Fig. 4.

Table 1

Integral REL (GeV cm^2 g^{-1}) Distribution of HZE-Particles in Apollo 16

Experiment	Detector	Fluence Φ(REL) [cm^{-2}]
Biostack I	CN	$17.3 \times REL^{-2.18}$
Biostack I	Lexan	$19.6 \times REL^{-2.18}$
MEED	CN + Lexan	$19.3 \times REL^{-2.26}$
Astronauts' dosimeters	Lexan	$70.1 \times REL^{-2.12}$

4. Observations on *Bacillus subtilis* Spores

The aim of the biological studies was to find out how a hit by a single HZE particle affects the development of a *Bacillus subtilis* spore. During incubation on nutrient broth agar at 37 °C the bacteria remained in fixed contact with the CN carrier sheet. They occupied the same position in the biological layer as they had during flight. Thus, during the whole development process, the hit spore or its descendant cell was identified by the aid of the etch cone in the CN sheet. Each hit spore was analysed individually. Its growth characteristics were compared with those of the surrounding cells, which were not hit by a heavy particle (these are the flight controls) and with those of the ground controls. The different stages of development studied were: germination, determined as change from bright to dark in phase contrast microscopy, outgrowth, determined as the formation of the first mature vegetative cell, and rate of cellular elongation of the first cell up to the fully grown vegetative cell.

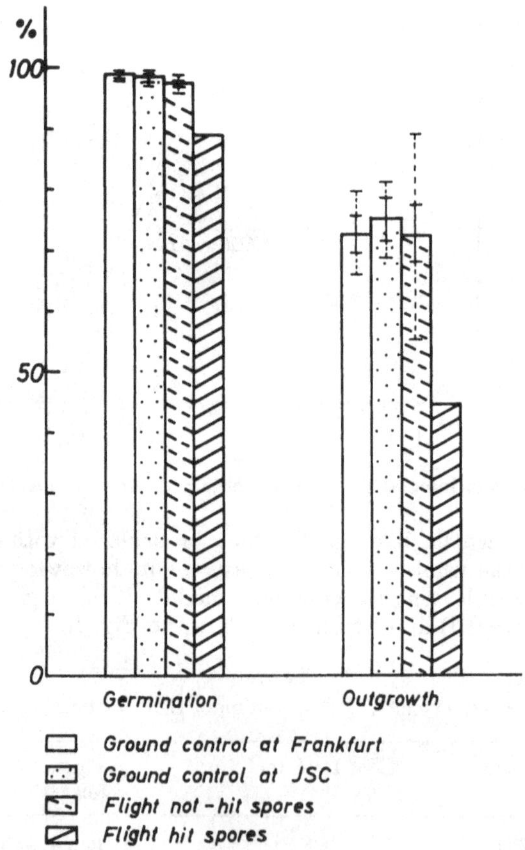

Fig. 5. Germination and outgrowth of *Bacillus subtilis* spores of the Biostack hit by an HZE particle and of the controls.

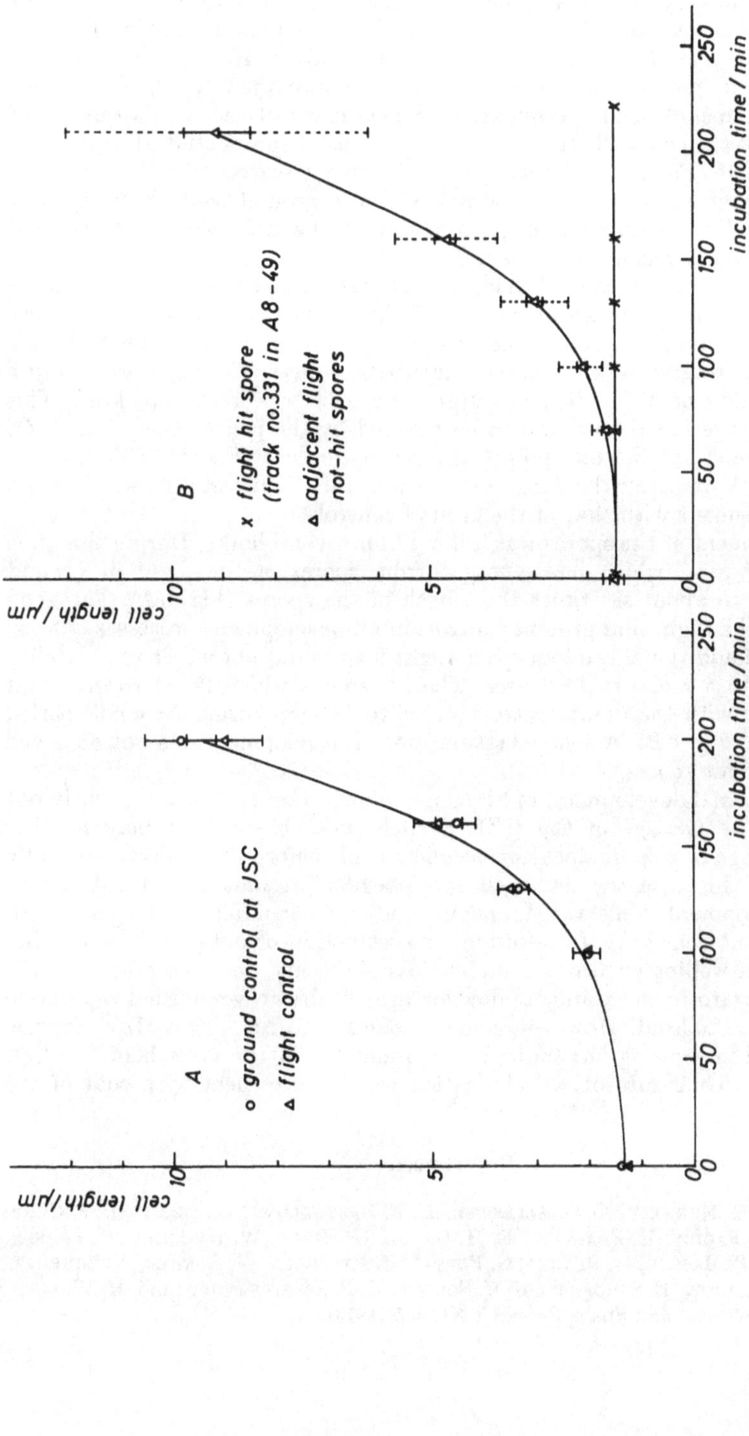

Fig. 6. Cellular elongation of cells emerged from spores of *Bacillus subtilis* flight and ground controls of the Biostack (A), and inhibition of growth of a spore hit by an HZE particle (B).

Germination was the first developmental stage. The phase darkening of the spores occurs because of a lowering of the refractive index, which is probably caused by excretion of spore constituents, slight swelling and redistribution of water within the spore [9]. This process has been proved to be highly radiation resistant. Irradiation with X-rays in doses of approximately 400 krad (which reduced the surviving fraction of colony formers to about 10^{-4}) did not influence the germination process. Much higher doses, approximately 2000 krad, are necessary to induce "pseudo-germination" which is correlated with an increased permeability of the cell wall [10]. About 90% of all hit spores studied so far were able to germinate (Fig. 5). This germinating fraction is scarcely smaller than that of the flight control, which latter agrees with the ground controls fairly well. Probably the slight reduction in germination of the hit spores is fortuitous. Pseudo-germination was not observed.

Outgrowth was the second developmental stage under investigation. During spore outgrowth the spore cases rupture and the embryo vegetative cell emerges, able to develop into a fully grown vegetative cell [11]. After irradiation with X-rays the outgrowing fraction decreases with irradiation dose. The dose producing a surviving fraction of 37% (D_{37}) of outgrowing cells was about 350 krad. This development stage has been found to bc affected by the penetration of an HZE particle. Only 45% of the hit spores achieved outgrowth (Fig. 5). This fraction is significantly lower than the 72% outgrowth of the flight controls, which again is in good agreement with that of the ground controls.

The development of the spores was followed for several hours. During this time the vegetative cells, which had emerged from spores of the flight or ground controls, grew to about six times the length of the spores (Fig. 6A). The good agreement of the flight and ground controls in all development processes studied shows that the additional random space flight factors did not affect the viability of the spores to a measurable degree. The hit spores which failed to grow out simultaneously with the flight controls ceased to develop during the whole period of observation (Fig. 6B). A delayed resumption of development was not observed in any of these hit spores.

However, normal development of hit spores whose outgrowth was evidently not inhibited by the passage of the HZE particle was observed. It may be that radiation damage of a spore does not become visible before the cell is nearly fully developed. In this case the damaged spore region presumably controls a late stage in development. This was found in studies on irradiation of spores with X-rays and deuterons [12]. In addition to a reduction of outgrowth, anomalies like monstrous swelling or nearly complete loss of opacity were observed. The D_{37} of X-rays in regard to the ability to develop into a fully grown normal vegetative cell was about 120 krad. However, none of our spores hit by an HZE particle that succeeded in growing out showed such anomalies in the growth of the first vegetative cell. Their rate of cell elongation was in agreement with that of the flight controls.

References

[1] H. Bücker, G. Horneck, O. C. Allkofer, K. P. Bartholomä, R. Beaujean, P. Cüer, W. Enge, R. Facius, H. François, E. H. Graul, G. Henig, W. Heinrich, R. Kaiser, H. Kühn, J. P. Massué, H. Planel, G. Portal, E. Reinholz, W. Rüther, W. Scheuermann, R. Schmitt, E. Schopper, J. U. Schott, J. P. Soleilhavoup and H. Wollenhaupt, Life Sciences and Space Research XI, 295 (1973).

[2] W. ENGE, K. P. BARTHOLOMÄ, R. BEAUJEAN, W. HEINRICH, K. FUKUI, G. HORNECK and H. BÜCKER, Proc. 8th Int. Conf. on Nuclear Photography and Solid State Track Detectors, Bucharest 1972.

[3] E. V. BENTON, Rep. USNRDL-TR-68-14, January 1968.

[4] R. BEAUJEAN and W. ENGE, Z. Phys. **256**, 416 (1972).

[5] W. ENGE, R. BEAUJEAN, K.-P. BARTHOLOMÄ and K. FUKUI, Life Sciences and Space Research XII, 51 (1974).

[6] E. V. BENTON and R. P. HENKE, Tech. Rep. No. AFWL-TR-72-5, October 1971.

[7] E. V. BENTON and R. P. HENKE, Tech. Rep. No. 24, University of San Francisco, Dept. of Physics, Sept. 1972.

[8] E. V. BENTON and R. P. HENKE, Techn. Rep. No. 20, University of San Francisco, Dept. of Physics, August 1972.

[9] G. W. GOULD, in: The Bacterial Spore, Academic Press, London and New York 1969 (p. 397).

[10] J. FARKAS and I. KISS, Acta Microbiol. Acad. Sci. Hung. **12**, 15 (1965).

[11] R. E. STRANGE and J. R. HUNTER, in: The Bacterial Spore, Academic Press, London and New York 1969 (p. 445).

[12] J. E. DONNELLAN and H. J. MOROWITZ, Radiation Res. **12**, 67 (1960).

Life Sciences and Space Research XII — Akademie-Verlag, Berlin 1974

STUDY OF COSMIC RAY EFFECTS ON *Artemia salina* EGGS DURING THE APOLLO 16 AND 17 FLIGHTS

H. PLANEL[a], J. P. SOLEILHAVOUP[a], Y. BLANQUET[a] and R. KAISER[b]

[a] Laboratoire de Biologie Médicale, Faculté de Médecine, Tolouse, France
[b] Laboratoire de Physique Corpusculaire, Centre d'Etudes Nucléaires de Strasbourg, France

We have used *Artemia salina* eggs, embedded in polyvinyl alcohol, to study the biological effects of heavy ions of cosmic rays. Each biological layer was sandwiched between track detectors. Hit eggs by heavy ions show a great inhibition of their developmental ability. A lower inhibition is observed for eggs that were flown but not hit. Simulation experiments are in progress to determine the factors responsible for inhibition of eggs that were not hit and to improve our knowledge of cellular damage induced by heavy ions.

1. Introduction

Biostack experiments (Dr. Bücker, principal investigator) were performed to study the biological effects of heavy ions of cosmic rays and the influence of other space factors, using a method which provides a good correlation between particle tracks and biological objects. This correlation was obtained by a combination of resting state biological objects and track detectors. In our laboratory we have studied the effects of cosmic radiation (heavy ions and background radiation) on the first stages of development of *Artemia salina* eggs.

2. Materials and Methods

In previous studies, the optimal conditions of culture and most suitable origin of eggs had been determined. Tetrawerke selected eggs (Tetrawerke, Melle, Germany) were used; these were tested before use. They were embedded in polyvinyl alcohol 20—30% aqueous solution. Thin layers, each including about five thousand eggs, were sandwiched with Ilford nuclear K2 and K5 emulsions.

After recovery, illumination of each *Artemia* layer and of each nuclear emulsion, covered by a millimetrie grid, allowed egg and grid shadows to be obtained. After processing, the trajectories of the heavy ions were geometrically reconstructed and the hit eggs were determined. With the help of nuclear emulsion scanning, hit eggs and eggs that were not hit were picked up with a scalpel after the polyvinyl alcohol had been cut around the eggs. Eggs were placed individually in crystal dishes with 2 ml of medium. The medium was prepared just before use as follows: doubly distilled water, 250ml; "Saltemia blue", 1 tablet; aquarium filtered water, 60 ml.

The medium was oxygenated with air for at least 6 hours. After the polyvinyl alcohol had been dissolved, the eggs were washed with the same medium and new medium was put in the culture dishes. Cultures were placed in an incubator at 27 ± 0.5 C and were illuminated by daylight. Cultures were checked every three hours from the 10th hour. Emergence (shells break open) occurs 10 to 12 hours after starting culture. Hatching (embryonic membranes disrupt) occurs generally 2 to 4 hours after emergence, giving rise to swimming nauplius larvae.

Investigations were carried out on 1200 eggs from Biostack I and on 300 eggs from Biostack II. We have studied the percentages of emergence, of hatching and of the number of live nauplii 4 to 5 days after hatching. These investigations have been performed on earth control eggs (embedded in PVA), on flown eggs that were not hit and on flown eggs hit by heavy ions. The total energy loss from heavy particles was determined by photometry by comparison with calibrated tracks.

3. Results

Biostack I. From 400 terrestrial control eggs 263 eggs emerged and the development of 17 emerged eggs was stopped. The others gave rise to nauplii; only 4 nauplii died in 4 to 5 days after hatching. For 400 flown eggs that were not hit 120 emergences were observed. After emergence, 21 eggs died and 7 nauplii died by 4—5 days. For 400 flown eggs that were hit by heavy ions only 38 eggs showed emergence. Nauplii were obtained from 25, and 6 nauplii died later. The percentages of emerged eggs, hatched eggs and swimming nauplii on the 4th—5th days of the experiment are shown in Fig. 1.

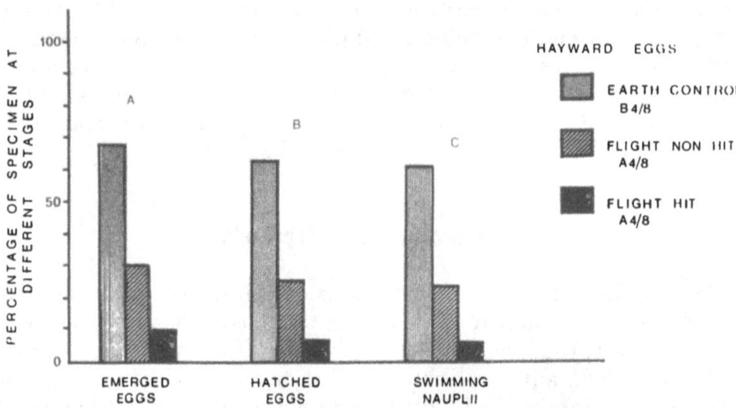

Fig. 1. Percentages of development of Artemia eggs.

Fig. 2 shows the percentages of egg emergences from the 18th hour to the 5th day. At the 18th hour, 20% of control eggs had emerged, while emergence was only starting for flown eggs.

Biostack II. Investigations, which are not completed, give percentages of emergence a little lower than those of Biostack I, and they are not exactly the same for each plate (Fig. 3). Nevertheless the results are similar to those of Biostack I.

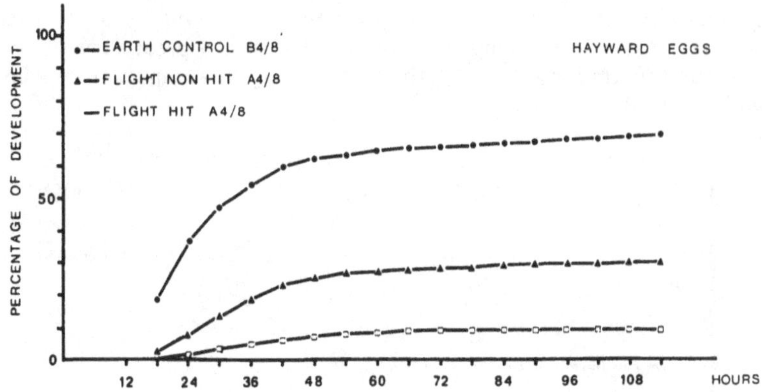

Fig. 2. Percentages of emergences as a function of time.

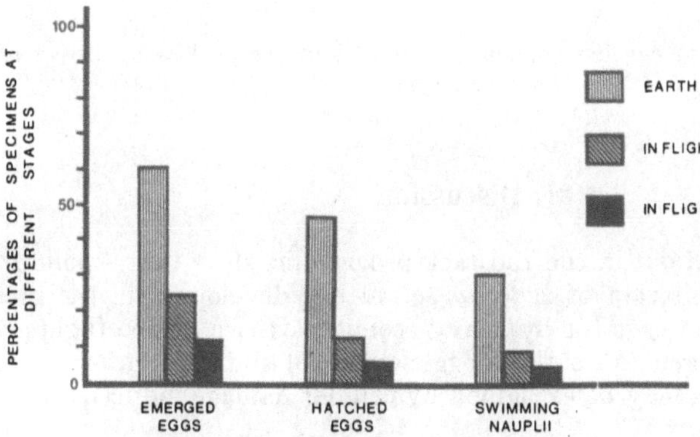

Fig. 3. Percentages of egg development in Biostack II.

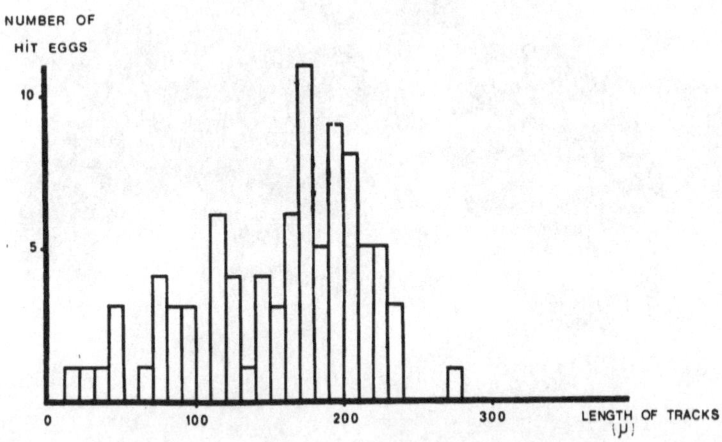

Fig. 4. Length of heavy ion tracks in Artemia eggs.

Energy loss was determined for 50 eggs hit by heavy ions, by photometry and after determination of track lengths. Most tracks cross a great part of the eggs: they are about 180 μm long (Fig. 4); the total energy loss ranged from some MeV to 130 MeV (Fig. 5).

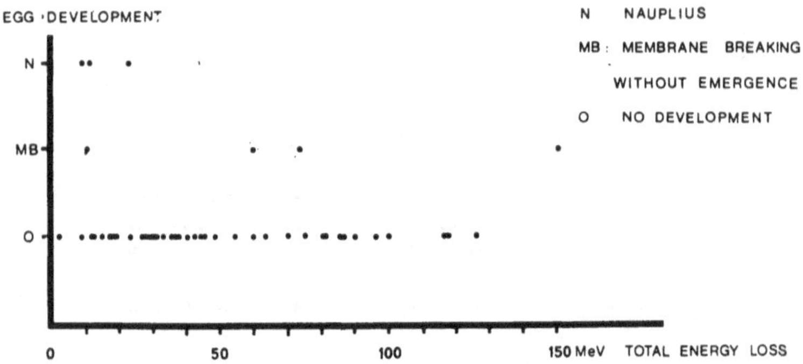

Fig. 5. Correlation between egg development capacity and energy loss by heavy ions in Artemia eggs.

4. Discussion

Investigations carried out in the Biostack programme show that Apollo flights result in a marked inhibition of *Artemia salina* egg development: the greatest inhibition is observed in eggs hit by heavy cosmic particles. Space flights result also in a delay of emergences for those eggs which could develop after recovery.

Inhibition of hit eggs may be explained by cellular damage induced by energy

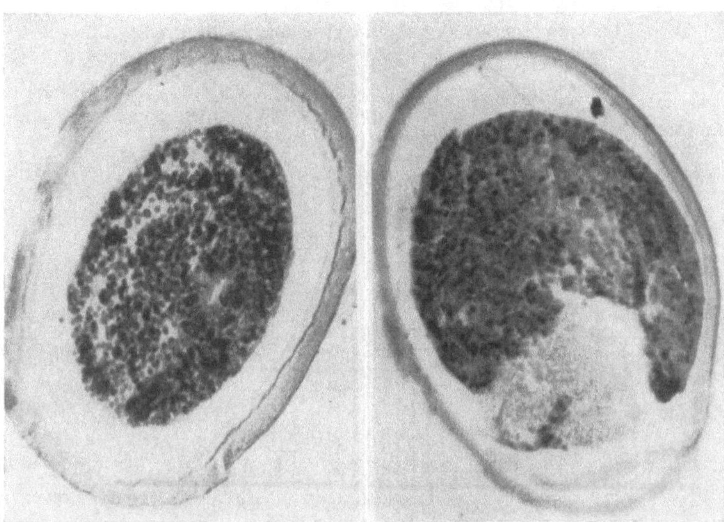

Fig. 6. Left, normal egg; right, egg irradiated by oxygen ions in Berkeley accelerator.

loss from heavy ions. Indeed, irradiation by 1or 2 oxygen ions of 1.2 GeV energy/nucleon in the Berkeley accelerator shows that a large part of the eggs, that is to say a great number of embryonic cells, is destroyed by the particles (Fig. 6). Eggs hit by ions in space flight which could still develop have only received 20 or 40 MeV energy/nucleon but it is possible that these eggs have been hit by tangential ions, so that only the capsule or subcapsular space was crossed by the particle.

For eggs that were not hit many factors may explain the inhibition of development: weightlessness, vibrations, accelerations, and cosmic background radiation (i.e. other components than heavy ions). Of course, these factors may act also on hit eggs and may have a synergistic effect.

Experiments are in progress in our laboratory to determine the effects of vibrations and accelerations. Preliminary experiments have been also performed to study the effect of background radiation, in spite of the very low doses given by these radiations. Eggs of the same age as those in Apollo 16, embedded in PVA, have been irradiated by ^{60}Co for 11 days, with a total dose of about 950 mrads. The eggs were studied in groups of 50 or individually. Under these experimental conditions, all the irradiated eggs have shown the same inhibition as seen in eggs which were not hit but were flown in the Apollo spacecraft. These results suggest that the inhibition of in-flight eggs may be due to the cosmic background. Nevertheless, new investigations must be carried out to confirm these preliminary results and to eliminate possible effects of other environmental factors.

Gravitational Biology

Life Sciences and Space Research XII — Akademie-Verlag, Berlin 1974

GRAVITY, WEIGHTLESSNESS AND THE GENETIC STRUCTURES OF ORGANISMS

N. P. DUBININ and E. N. VAULINA

USSR Academy of Sciences, Moscow, USSR

The whole evolution of life on earth has proceeded under the action of earth's gravity which must have influenced the structure and function of organisms. During space flights organisms are exposed to the entirely new condition of weightlessness, and to variations in gravity that produce various changes. Current flight data suggest that organisms most often respond to weightlessness by disturbances in physiological function, which are reversible after the brief exposures that have at present been possible. Only longer space flights will show whether these changes will be compensated, or will lead to alterations in the more important systems of organisms including their hereditary properties. There is evidence that weightlessness has a direct effect on genetic properties. Thus, in microspores of *Tradescantia* abnormal mitoses (that were not observed in the ground-based control) were recorded to an extent of 3%. The numerous changes in various vital systems suggest that for terrestrial organisms weightlessness is a factor which, if administered for a long time, may bring about serious disturbances in their activities and heredity. The higher the evolutionary position of the organism, the more pronounced the effect is likely to be.

The gravitational field is one of the basic characteristics of our planet. Space flights, missions to the moon and expeditions to other planets of the solar system introduce organisms into an environment with a different gravitational field. When the system is moving with an acceleration equal to the gravitational force of the earth, a state of weightlessness is created, and material particles inside the system exert no pressure on each other. Weightlessness is a completely new condition for terrestrial organisms, since their whole evolution has proceeded under the action of gravity. In the course of their evolution all organisms including man have developed a number of important structural and functional peculiarities as adaptations to gravity. Zero gravity is one of the factors that are extremely difficult to investigate experimentally for it cannot be simulated on earth. Nevertheless, interesting evidence is accumulating for the role of gravity in the evolution of organisms and their orientation and development, and for the effects of gravity on heredity. It has been shown that gravitational forces may largely determine the form and the size of organisms, the development of supporting skeletal structures and their energetics, thus producing marked changes in organs and their functions [1—13]. A chain of evolutionary adaptations to gravity can be traced in organisms after they came to live on the land instead of the sea where gravitational forces were largely balanced by the buoyancy of the liquid. This imposed a number of new properties on vertebrate animals, such as reduced body size, relative larger skeleton, development of limbs, increased heart weight owing to the increased expenditure of energy on the land, certain peculiarities of hemoglobin synthesis and prolongation of the life span of red blood corpuscles; these illustrate the complex adjustments needed [1, 2, 7, 8, 10, 13, 14].

In plants there were other changes, such as reduction of size in herbaceous plants, development of powerful supporting structures in trees, elaboration of conducting and skeletal tissue systems, development of complex adjustments to maintain the supply of water, oxygen and mineral salts and protection from light and heat [12, 15]. General adaptations of organisms to gravity provides for their stable spatial position with reference to the gravitational field, evolution of supporting structures determining the form of the organism and position of organs, and energetics and biochemistry to enable the organism to function in the gravitational field.

There are no organisms on earth not subjected to gravity, for every organism develops some specialized organ of orientation. In unicellular protozoa this function is mediated by certain organelles. Amyloplasts have been shown to function as statoliths and the distribution of mitochondria inside the cell depends on gravitational forces [16]. Infusoria have an organelle shaped like a vacuole filled with otoconium; in *Paramecium* food vacuoles serve the same purpose [17—20]. More complicated organisms can be divided into two groups according to their mode of orientation: mobile organisms (animals) or immobile organisms (plants).

In animals, the organ of static equilibrium evolved from a statocyte to a perfect organ in mammals consisting of an intricate maze of ampullae, a utriculus, sacculus and organ of Corti. The evolution of animals has resulted in the emergence of a sensory and motor system determining the position of the whole body and its parts with respect to the gravitational field and able to sense deviations from the normal position. The system corrects the position of the body, retaining it within vitally important limits, and it also keeps a record of the amount and direction of linear acceleration involved in movement of animals [21].

Plants have subcellular "sensors of gravity", statoliths, which are displaced as a result of reorientation of the organ with reference to the gravitational field, and these mediate the response of the cell or organ. In rhizoid apical cells of the giant unicellular alga *Chara*, statoliths control the growth of cell walls, either accelerating or inhibiting the flow of substances making up the cellular membrane [22]. In multicellular plants statoliths are identified with amyloplasts [7]. Their movements within the cell under the action of gravity create polarity in the organ, which may then change in its growth. It has been suggested that the Golgi apparatus is a cell geosensor [16]. In plants orientation to gravitational field is shown by orthotropic (vertically arranged) and plagiotropic (horizontally arranged) organs. Plants normally display negative geotropism in stems and positive geotropism in roots. The mechanism was investigated by N. G. Holodny [23] and Went [24] who put forward a hormonal theory of tropisms. Many workers are now studying tropisms and their role in the life of plants. It has been shown that in geotropic stimulation an indole compound of the indole acetic acid (IAA) type is formed and that it is the effector in geotropic reactions [27—28]. IAA is responsible for the bioelectric polarization of tissues in the geotropic action in growth [29, 30]. Distinct transverse polarization of energy metabolism was found in geotropic stimulation of orthotropic organs [31, 32] though the total was constant. Protein synthesis increased in one half of the axial organs as a result of an accelerated growth of cells and cell walls in geotropic curves [33]. It is known that gravitational forces cause certain spatial orientation of both orthotropic and plagiotropic organs

[34—35]. It has also been shown by a number of workers that the character of geotropic reactions in plant organs is strongly dependent on oxidation-reduction systems [36, 37].

In vertebrates the blood circulation system evolved to meet the demands of energy supply. It has been shown in chicks, rat, mice and hamsters that chronic centrifugation causes a retardation of growth, and reduction of size, body weight and fat content [38]. In other words, the organism responds to increased gravity with an increased energy demand.

All the foregoing concerns gravitational effects on terrestrial organisms without direct investigation of effects of weightlessness. Space flight experiments permit us to understand the above-mentioned facts. They are not numerous but are of undoubted interest. Most of the reliable evidence so far available is on the physiology of man, for in numerous, though short, space flights various physiological characteristics of man were measured. These date have been widely discussed in the literature. They support our understanding of the gravitational effects arrived at by our analysis of evolution of terrestrial organisms. During space flights when there is no gravitational field to allow orientation in the environment, disorders of the sensory-motor apparatus are marked and are seen as disturbances of the vestibular apparatus, coordination, neuro-muscular control and control of the upright posture etc. [39—43]. Disturbances in the organs of support and locomotion are also marked. Changes in the water-salt metabolism affect significantly the mineral part of bones, resulting in thinning of skeletal bones. The tone and strength of the lower extremities decrease. These changes tend to be more pronounced with increasing duration of space flight [41, 43]. In the decreased gravity on the moon a reduction is found in the elasticity of muscles active in running. The mode of locomotion is also changed, i.e. it is most convenient to jump on two feet like a kangaroo on the moon at $^1/_6$ g, and to "swim" in space or on board a space vehicle [43]. Noticeable differences were observed in energetics and related vital systems, e.g. respiratory organs. Thus, V. N. Tereshkova and V. F. Bykovsky showed a 30% increase in the oxygen absorbing capacity of the lungs persisting for 8—10 hours after the flight [44]. Among other changes observed were a decrease in the blood plasma volume, red blood corpuscle volume and total blood volume [39—42, 45], and a number of disturbances in the cardio-vascular and endocrine systems. Despite the fact that with the present duration of space flights all physiological disorders are reversible [42], they are so extensive that they are a cause for concern when an attempt is made to forestall the consequences of longer space flights.

Responses of the genetic systems and of pertinent organs and organelle, to space flight factors (including weightlessness) are studied on animal and plant objects which are chosen for several traits, namely, their genetics are well known, they are of modest size and consume little energy (when experiments are devised certain restrictions are imposed on weight, size and consumption of energy) and their response to space agencies are readily recorded. Objects at different levels of organization from isolated DNA molecules to mammals have been chosen, namely viruses, micro-organisms, unicellular algae and paramecia, seeds and adult plants, insects (*Drosophila*, *Tribolium* and *Habrobracon*), frog eggs and mice.

It should be noted that the above-mentioned objects displayed quite a number of physiological and biochemical changes. In many experiments on weightlessness disturbances of metabolism, giving rise to disorders in physiological functions, were caused by changes in the gaseous, hydrodynamic, thermal and mass metabolism of the organism. The higher the evolutionary position of the organism, the more serious are the physiological disturbances. Many of those shown by man were discussed earlier. Changes of the same order were observed in animals [46—48]. A number of physiological changes as a result of metabolic abnormalities were also seen in plants, e.g., it was shown with wheat seedlings that respiration increased, the orientation of organs changed (presumably due to displacement of auxins with their concomitant reactions), the volume of interphase nuclei in the cells of primary and lateral seedling rootlets increased and their cells elongated markedly (by 13.5%) [49]. Redistribution of cellular organelles caused by weightlessness were also observed. Most pronounced changes were produced in starch grains, which act as gravity receptors. In weightlessness they were distributed throughout

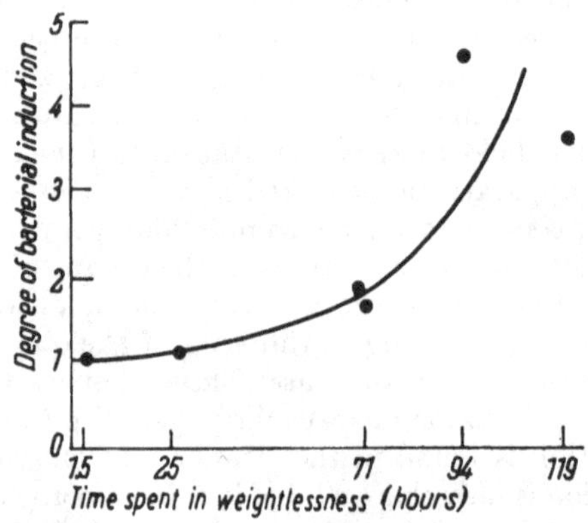

Fig. 1. The ratio of the number of phage particles found in samples of lysogenic bacteria *E. coli* K-12 (λ) exposed on space ships Vostok to the number of phage particles in the control (spontaneous level of the phage production) [57].

the cell, while in the ground-based control they were grouped about the lower cell wall [49]. An increase of the life span of the *Drosophila* females exposed to space [50] and a decrease in the activity of xanthine dehydrogenase in *Habrobracon* [51] as well as reduction of the growth rate in *Amoeba* [52] are all evidence for metabolic changes. An acceleration of growth in bacteria (*Salmonella typhimurium*) [53] also seems to be the result of changes in the cell energetics caused by weightlessness [54]. The authors suggest that the bacterial cells are randomly distributed in the population under weightless conditions. Under these conditions nutrients are also distributed evenly through the medium; therefore cell energy is used for reproduction rather than for migration toward regions with better nutrients. Tobias [54] believes that weightlessness, affecting metabolism in organisms, retards the rate of complicated genetical changes. Flight experiments of longer duration will show to what extent this effect on organisms will be balanced by metabolic changes.

Some of the results obtained indicate that there are indirect effects of weightlessness on genetic systems acting through metabolic changes. An after-flight increase in the phage production of a lysogenic culture of *Escherichia coli* K-12 (λ) [55] (Fig. 1), which correlates with the duration of flight, presumably represents an indirect effect of weightlessness. The induced value was appreciably above the level expected from the dose of ionizing radiation received in flight. Vibration can hardly contribute to this effect since it has been shown that vibration as such does not induce phage production. Yet when interacting with radiation, vibration can augment the production of phage provided the vibration is exercised both before and after irradiation. Vibration had no effect on phage production induced by ionizing radiation, if the time between these agencies was considerable (1.5 hours in the experiment reported).

A fall in the after-flight survival of a resting *Chlorella* culture (Fig. 2) correlated with the duration of flight and this may also be accounted for by disturbances

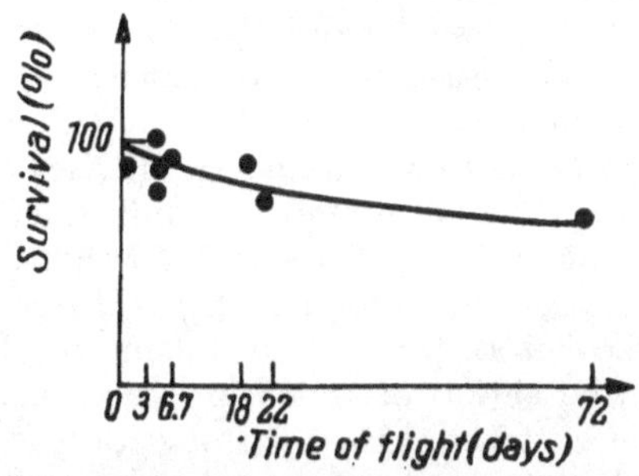

Fig. 2. Survival of *Chlorella* cells at various times of flight. The earth control should be taken as 100%.

in the metabolic processes under weightless conditions [56—59]. The possibility of such disturbances was shown in preliminary experiments which demonstrated that survival and mutability of the *Chlorella* culture depended upon the position of the cells with reference to the gravitational field [60].

Wheat seedlings exposed on board the space ship had a lowered mitotic index [49]. Examination of the cells at different mitotic phases revealed that the phase most sensitive to weightlessness was early prophase, because among cells of primary and lateral root tips exposed to weightlessness those in early prophase were appreciably fewer in number than in the control.

Before describing the effect of weightlessness on genetic systems we shall discuss some theory. As early as 1940 V. I. Vernadsky in his speculations on the effect of gravitational forces on organisms assumed that organisms which are on the verge of being living (comparable in size with molecules) are not affected by gravity but are wholly under the action of the molecular forces [61]. According to Pollard's calculations [62], weightlessness can affect the distribution of intracellular components only when their potential energy is greater than that of Brownian movement. Making use of Pollard's method Kondo predicted the possibility of direct effects of weightlessness on cellular genetic structures of organisms of various levels of organization [63]. He presumes that weightlessness will not act on the genetic structures of micro-organisms, particularly chromosomes of *Escherichia coli* and haploid yeasts. *Drosophila* chromosomes will be on the verge of showing such effects. Chromosomes in *Tradescantia* and man will be affected by weightlessness. According to Kondo, in higher organisms gravitational forces play an important role during cell division when the density of chromosomes varies. In his opinion weightlessness will undoubtedly have an indirect effect on genetic systems. In micro-organisms it will be shown by disturbances of intercellular relations in large populations, in higher forms by abnormal interactions between subcellular particles.

The findings of the experiments using mutagenic treatment of organisms before and after space flight are the most interesting of the data so far available on the effect of weightlessness on heredity. The cells exhibited effects caused by weightlessness shown as changes in the frequency of mutations and in the sensitivity of chromosomes to mutagens. The effect of the pre-flight treatment indicates that weightlessness inhibits repair systems; the effect of the post-flight treatment suggests that weightlessness gives rise to potential genetic lesions.

Dubinina and Chernikova [64] report that after exposure on board the space vehicle *Crepis capillaris* seeds were more than twice as sensitive to ethyleneimine. A change in the radioresistance of organisms was demonstrated in dry plant seeds and *Chlorella* cultures using pre- and post-flight exposure to gamma-rays [59, 65, 66]. *Drosophila* males which underwent X-ray treatment prior to the flight showed a higher frequency of recessive sex-linked lethals and translocations than the ground-based controls [67]. The effect of weightlessness may perhaps be pronounced at the population level, though this still remains speculative. However, Zhukov-Verezhnikov et al. [55] obtained some results on *E. coli* K-12 (λ) which point to a selective action of weightlessness, i.e. after double exposure to space flight lysogenic bacteria showed a statistically significant fall in phage production (1.74 ratio of experiment to control) compared with a culture exposed only once (2.3).

A direct effect of weightlessness was demonstrated by N. L. Delone on chromosomes of *Tradescantia* microspores exposed on board the satellites Vostok 3–6 and Voskhod [68, 69]. Abnormal mitoses that were not observed in the control were recorded at a level of 3%. These disturbances were expressed as altered orientation of chromosomes (the third type of abnormality in the system of scoring), delay of non-disjunctional chromosomes (the fourth type), multi-polar mitoses (fifth type), non-disjunction of chromosomal complexes (first and second types) (see Fig. 3). The investigators were faced with considerable difficulties in the analysis (because of the complexity of the experimental conditions, namely temperature variations from 20° to 30 °C because these influence the duration of mitosis and affect the sensitivity of the different mitotic phases to various agencies. Nevertheless the authors unambiguously demonstrated (particularly in those experiments using fixation during the flight) the dependence of the effect of weightlessness upon the duration of exposure (Fig. 4). Disorders in the mitotic spindle of microspores, megaspores and root tips in *Tradescantia* were also found by American workers [70].

The evidence discussed above is sufficient evidence for both direct and indirect effects of weightlessness on genetic systems. In addition, weightlessness has an appreciable modifying role on other space flight factors. In experiments with gamma irradiation at various doses on board the Biosatellite 2 both synergistic and antagonistic effects were observed. In *Tribolium* irradiated in weightless conditions the number of females with abnormalities of wings and the frequency of dominant lethals in females increased [71], which, in the author's opinion, was the result of less effective repair systems under conditions of weightlessness. Under the same condition the number of *Drosophila* with deformed thoraces and abnormal wings increased. In addition, the flies hatched from eggs laid in space showed an increase in the frequency of chromosomal breaks and translocations

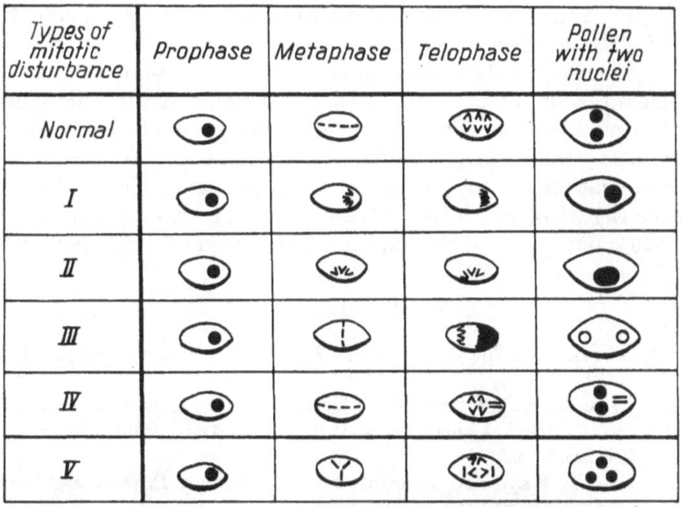

Fig. 3. A scheme of types of mitotic disturbances in *Tradescantia paludosa* microspores [68, 69].

between chromosomes II, III and V, and in the amount of sex-linked recessive lethals [50, 67]. In *Tradescantia* irradiated under weightless conditions there was an increase in the proportion of abortive pollen, the frequency of pollen micronuclei and the amount of stamen hairs with inhibited growth [70]. An antagonistic effect was discovered in *Drosophila* irradiated in space; the flies showed a reduced loss of X-chromosome markers compared with the ground-based control [67]. In *Neurospora*, analysis of suspensions showed a decrease in the mutation frequency in flight compared with the ground-based control, because of a decreased number of point mutations; changes were observed in the number of chromosome deletions [72]. Vibration decreased the number of radiation-induced translocations in *Drosophila* [67], increased the frequency of spontaneous recessive lethals in

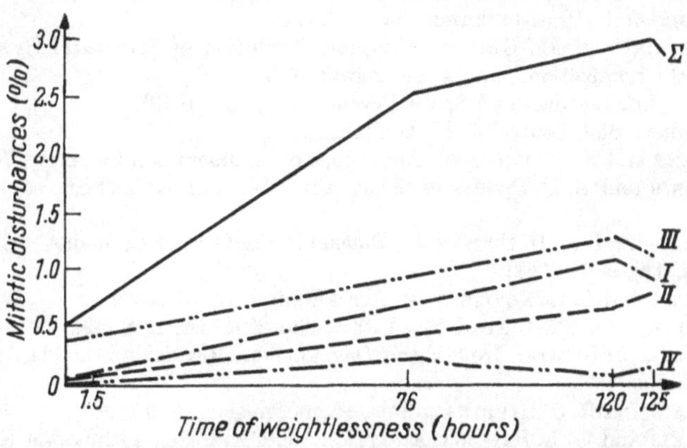

Fig. 4. Percentage of different types of mitotic disturbances in *Tradescantia paludosa* microspores [70]. Σ, total percentage of mitotic disturbances; I-IV, percentage of different types of mitotic disturbances.

Habrobracon [50] and recessive and dominant lethals in *Drosophila* [73] and the emergence of *Artemia* from cysts [51].

The flight experiments described above have shown that weightlessness gives rise to disturbances, or creates conditions under which disturbances in the genetic structures can readily arise. The evidence for the physiological, biochemical and genetic effects of gravity suggests that advances in gravitational biology have not only an educational value but also have practical implications for long-term manned space flights of the future. Overcoming the barriers imposed by weightlessness presents one of the most important challenges for all space sciences.

References

[1] K. E. TSYOLKOVSKY, in: The Way to Stars, Publ. House of the USSR Acad. Sci., Moscow 1961.
[2] A. N. SEVERTZOV, Main Directions of the Evolutionary Process, Biomedgiz, Moscow 1934.
[3] A. N. SEVERTZOV, Morphological Laws of the Evolution, Publ. House of the USSR Acad. Sci., Moscow 1939.
[4] I. I. SHMALGAUSEN, Ways and Laws of the Evolutionary Process, Publ. House of the USSR Acad. Sci., Moscow 1940.
[5] I. I. SHMALGAUSEN, The Origin of the Land Vertebrates, Nauka, Moscow 1964.
[6] H. SZARSKI, Pochodzenie Plazow, Warszawa 1961.
[7] D. THOMPSON, Growth and Form, Cambridge Univ. Press, Cambridge 1917.
[8] V. YA. BROVAR, The Gravitational Force and Animal Morphology, Publ. House of the USSR Acad. Sci., Moscow 1960.
[9] V. S. IVLEV, J. Gen. Biol. **20**, 2 (1959).
[10] P. A. KORSHUEV, Evolution of Gravity, Weightlessness, Nauka, Moscow 1971.
[11] P. A. KORSHUEV, Aviation and Space Medicine, Moscow 1963.
[12] V. F. RAZDORSKY, Plant Architectonics, Soviet Science, Moscow 1955.
[13] L. A. ZENKEVICH, J. Gen. Biol. **5**, 3 (1944).
[14] G. FOXON, Bul. Rev. **30**, No. 2 (1955).
[15] G. H. MOLOTKOVSKY, Dokl. Akad. Nauk SSSR **78**, 3 (1951).
[16] J. SHEN-MILLER, unpublished.
[17] O. KOEHLER, Arch. Protistenkunde, **45**, 37 (1922).
[18] O. KOEHLER, in: Handbuch Norm. Pathol. Physiol., II, Berlin, 1027 (1926).
[19] Y. DEMBROWSKI, Arch. Protistenkunde, **66**, 104 (1929).
[20] H. MERTON, Arch. Protistenkunde, **85**, 33 (1935).
[21] YA. A. VINNIKOV et al., Gravity Receptor. Evolution of Structural Cytochemical and Functional Organization, Nauka, Leningrad 1971.
[22] P. LARSEN, Life Sciences and Space Research XI, 141 (1973).
[23] N. CHOLODNY, Biol. Zentralbl. **47**, 604 (1927).
[24] F. W. WENT and K. V. THIMANN, Phytohormones, Macmillan Co., New York 1937.
[25] A. I. MERKIS and A. D. PUTRIMAS, Proc. Acad. Sci. Lithuanian SSR, ser. B, **1** (42), 101 (1967).
[26] A. I. MERKIS and A. D. PUTRIMAS, Biochem. Conf. of Belorussian SSR and Baltic Republics, Thesis, **I** (1968).
[27] A. I. MERKIS and L. L. NOVITSKENE, Agrokhimiya, **10**, 87 (1969).
[28] A. I. MERKIS et al., Proc. Acad. Sci. Lithuanian SSR, ser. B, **2** (49), 65 (1969).
[29] A. I. MERKIS, in: General Growth and Development Regularities in Plants. Mintis Publ. House, Vilnius 1965.
[30] A. I. MERKIS and R. S. LAURINAVICHUS, Plant Physiol. **15**, 5 (1968).
[31] A. I. MERKIS and L. L. PRIALGAUSKAITE, Proc. Acad. Sci. Lithuanian SSR, ser. B, **3** (32), 103 (1963).
[32] A. I. MERKIS and O. YU. RUPAINENE, Proc. Acad. Sci. Lithuanian SSR, ser. B, **1** (36), 97 (1965).

[33] A. I. Merkis and A. S. Marchiukaitis, Proc. Acad. Sci. Lithuanian SSR, ser. B, **3**, (38), 59 (1965).
[34] B. M. Katunsky, Dokl. Akad. Nauk SSSR, **3** (12), 7 (1936).
[35] N. V. Bochurova, Plant Physiol. **17**, 3 (1970).
[36] N. S. Turkova, Kazakh. Affill. Publ. House USSR Acad. Sci., ser. Plant Physiol. and Biochem., 1 (1945).
[37] N. S. Turkova, Bull. Moscow. Natur. Society, Sect. Biol., **57**, 4 (1952).
[38] G. C. Pitts, Life Sciences and Space Research XI, 171 (1973).
[39] C. A. Berry, Life Sciences and Space Research XI, 89 (1973).
[40] C. A. Berry et al., NASA, SP-121, Washington, D.C. (1966).
[41] N. N. Gurovsky et al., Life Sciences and Space Research XI, 77 (1973).
[42] M. A. Cherepakhin et al., Life Sciences and Space Research XI, 117 (1973).
[43] R. Margaria, Life Sciences and Space Research XI, 177 (1973).
[44] V. V. Parin, paper presented at Space Science Symp., Florence 1964.
[45] C. L. Fischer et al., J. Amer. Med. Assoc. **200**, 99 (1967).
[46] E. H. Kass, in discussion at XVth COSPAR Meeting, in Madrid 1972.
[47] J. Hanley et al., Life Sciences and Space Research XI, 123 (1973).
[48] N. A. Gaidamakin et al., Kosm. Issled. **7**, 931 (1969).
[49] S. W. Gray and B. F. Edwards, in: The experiments of Biosatellite II, NASA, Washington, D.C., 1971 (p. 123).
[50] I. I. Oster, in: The experiments of Biosatellite II, NASA, Washington, D.C., 1971 (p. 41).
[51] R. C. von Borstel et al., in: The experiments of Biosatellite II, NASA, Washington, D.C., 1971 (p. 17).
[52] J. H. Abel et al., in: The experiments of Biosatellite II, NASA, Washington, D.C., 1971 (p. 291).
[53] R. H. T. Mattoni et al., in: The experiments of Biosatellite II, NASA, Washington, D.C., 1971 (p. 309).
[54] C. A. Tobias, Life Sciences and Space Research XI, 127 (1973).
[55] N. N. Zhukov-Verezhnikov et al., in: Problems of Space Biology, Nauka, Moscow 1965 (p. 261).
[56] E. N. Vaulina et al., Kosm. Issled. **5**, 2 (1967).
[57] E. N. Vaulina and I. D. Anikeeva, Japan. J. Genetics, **43**, 469 (1968).
[58] E. N. Vaulina et al., Life Sciences and Space Research IX, 105 (1971).
[59] N. P. Dubinin et al., Life Sciences and Space Research XI, 105 (1973).
[60] E. N. Vaulina and I. D. Anikeeva, paper presented at XVIth COSPAR Meeting, Konstanz 1973.
[61] V. I. Vernadsky, Biogeochem. Sketches, Publ. House USSR Acad. Sci., Moscow, Leningrad 1940.
[62] E. C. Pollard, J. Theoret. Biol. **8**, 113 (1965).
[63] S. Kondo, Japan. J. Genetics **43**, 468 (1968).
[64] L. G. Dubinina and O. P. Chernikova, Japan. J. Genetics **43**, 470 (1968).
[65] Yu. V. Farber et al., Space Biol. and Med. **5**, 24 (1971).
[66] K. P. Garina and N. I. Romanova, Kosm. Issled. **8**, 158 (1970).
[67] L. S. Browning, in: The experiments of Biosatellite II, NASA, Washington, D.C., 1971 (p. 55).
[68] V. V. Antipov et al., in: Problems of Space Biol. **4**, Nauka, Moscow 1965 (p. 248).
[69] N. L. Delone and B. B. Egorov, Dokl. Akad. Nauk SSSR **166**, 713 (1966).
[70] A. H. Sparrow et al., in: The experiments of Biosatellite II, NASA, Washington, D.C., 1971 (p. 99).
[71] B. Buckhold et al., in: The experiments of Biosatellite II, NASA, Washington, D.C., 1971 (p. 79).
[72] F. J. de Serres et al., in: The experiments of Biosatellite II, NASA, Washington, D.C., 1971 (p. 325).
[73] Ya. L. Glembotsky and G. P. Parfenov, in: Problems of Space Biology, Nauka, Moscow 1962 (p. 98).

Life Sciences and Space Research XII — Akademie-Verlag, Berlin 1974

HAEMODYNAMIC CHANGES CAUSED IN RATS BY PROLONGED ACCELERATIONS

M. Wojtkowiak

Institute of Aviation Medicine, Warsaw, Poland

Experiments were performed on 60 rats undergoing about $5G_z$ acceleration. The animals were killed in liquid nitrogen at various times after centrifugation (time of centrifugation 1 hour). Before centrifugation ^{131}I-labelled albumin was injected intravenously and after the experiment radioactivity of muscles from the extremities was measured.

The following conclusions were drawn from the experiments: (i) About $5G_z$ acceleration in the longitudinal axis results in blood displacement to the hind limbs with extravascular leakage of albumin. (ii) Immediately after centrifugation lowering of blood amount in the hind limbs occurred. During the first 15 minutes this decrease progresses very quickly, while later it is much smaller. Nevertheless even 3 hours after centrifugation there was only partial return to normality. (iii) Transudation of albumin in the hind limbs lasts a few hours after centrifugation. (iv) The haemodynamic changes caused by centrifugation consist of blood stasis and extravascular leakage of proteins.

1. Introduction

Acceleration of about $5G_z$ for one hour results in haemodynamic changes, i.e. blood displacement and extravascular leakage of albumin [1—4]. It seems [2] that under these conditions only small amounts of globulin penetrate the vascular walls. Morphological changes in the vascular walls were also observed [1—3, 5]. There is displacement of blood to the hind extremities, blood stasis and hypoxia in a large part of the body, transudation of blood proteins, mainly albumin, impairment of the vascular wall and extravascular leakage of blood. In human beings displacement of blood to the lower extremities, albuminuria, haematuria and somtetimes even petechiae were observed after several minutes of centrifugation at $\sim 5G_z$, the degree of these changes being dependent on individual tolerance. In rats, during prolonged centrifugation ($\sim 3G_z$ 3 hours daily during 3 weeks) [5, 6] increased permeability of blood vessels and morphological changes in blood vessels in kidneys and hind limbs were seen. It seems that the mechanism of these changes is similar both in humans and in experimental animals. It is therefore interesting to evaluate the reversibility of this increase in vascular permeability for proteins and the period required for return to normal.

2. Material and Methods

The experiments were performed on 60 albino Wistar rats, weighing about 200 g each. The animals were divided into six groups — one control group and five groups of animals exposed to $\sim 5 G_z$ acceleration on an animal centrifuge for one hour. The animals of one experimental group were killed in liquid nitrogen during centrifugation [7]. The rest were killed at various times after centrifugation (15 min, 1, 2 and 3 hours). No lethal effect of centrifugation was seen during the experiments, nor any macroscopic changes in internal organs.

All the animals were injected intravenously before centrifugation with $0.25 \mu Ci\, g^{-1}$ body weight of ^{131}I-labelled human blood albumin from the Institute of Nuclear Investigations, Warsaw, Poland. After decapitation samples (about 1 g) of muscles from fore and hind limbs were taken, dried, weighed, homogenized in 1 N KOH at 60 °C and the radioactivity was measured in a scintillation counter average of 3 determinations).

3. Results

The results are summarized in Table 1 and in Fig. 1. All the results are presented as mean values and percentage ratios of the radioactivity. In animals frozen with (iquid nitrogen during centrifugation a much higher amount of ^{131}I-albumin was

Table 1

Radioactivity of Muscles at various times after Centrifugation

Group of rats	Radioactivity of muscles of the limbs (counts/min/mg dry weight)* Fore limb	Hind limb	Percentage ratio hind/fore
Control	54.7 ± 11.0	53.3 ± 9.3^{b}	97.4
Frozen during centrifugation	49.8 ± 6.1^{a}	73.5 ± 7.3^{a}	143.7
15 minutes after centrifugation	49.0 ± 6.0^{a}	58.9 ± 5.2^{ab}	120.3
1 hour after centrifugation	59.2 ± 5.2^{ab}	66.6 ± 6.5^{ab}	112.5
2 hours after centrifugation	54.5 ± 3.6	59.8 ± 2.9^{ab}	109.7
3 hours after centrifugation	56.2 ± 4.3	58.1 ± 4.7^{b}	103.4

* mean and standard deviation

[a] statistically significant (Student's test) compared with control values ($p < 0.05$);
[b] statistically significant (Student's test) compared with animals frozen during centrifugation ($p < 0.05$).

The percentage ratio was not analysed statistically.

found in the hind limbs (143.7%) compared with the radioactivity of muscles from the fore limbs. After centrifugation the values returned to normal rapidly in the first 15 minutes and much more slowly thereafter (Fig. 1).

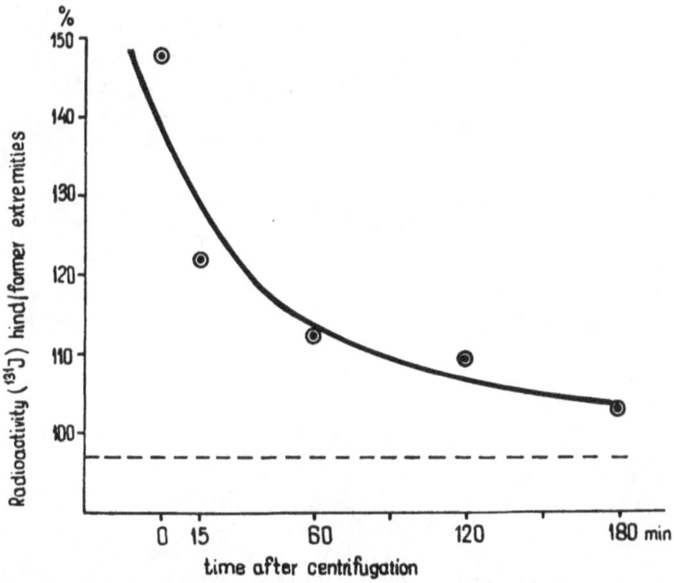

Fig. 1. Ratio of ^{131}I-albumin in muscle of hind limbs to fore limbs after centrifugation.

4. Discussion

Our earlier results [3, 4], as well as experiments performed by other authors [8—11], showed displacement of blood to hind limbs with increase of volume of the hind limbs. No doubt blood stasis and transudation of blood proteins play a significant role in increasing the limb volume as erythrocytes do not penetrate the vascular wall at accelerations of $9G_z$ or less for short periods. The rapid decrease of radioactivity is probably due to stasis of blood in the vessels and rapid return of the blood circulation to normal after centrifugation.

References

[1] S. Barański et al., Post. Astron. **4**, 5 (1969) (in Polish).
[2] P. Czerski et al., Patol. Pol. **29**, 395 (1968) (in Polish).
[3] P. Czerski et al., Med. Lotn. **21/22**, 139 (1967) (in Polish).
[4] J. S. Garrow, A Textbook of Aviation Physiology, Churchill, London 1965 (p. 589).
[5] P. Czerski et al., Post. Astron. **4**, 113 (1969) (in Polish).
[6] P. Czerski et al., Post. Astron. **6**, 126 (1971) (in Polish).
[7] C. F. Gell and D. Cranmore, J. Aviat. Med. **27**, 497 (1956).
[8] J. P. Henry et al., Feder. Proc. **8**, 73 (1949).
[9] J. P. Henry et al., Am. J. Physiol. **159**, 573 (1949).
[10] P. Howard, J. Physiol. **9**, 49 (1959).
[11] R. Senelar, in: Bioassay Techniques for Human Centrifuge, Pergamon Press, London, 1961 (p. 107).
[12] M. Wojtkowiak et al., Wynal. Racjonal. Wojsk. **18**, 22 (1966) (in Polish).

Life Sciences and Space Research XII — Akademie-Verlag, Berlin 1974

EFFECT OF HYPERGRAVITY AND HYPERTHERMIA ON ANTIDIURETIC HORMONE SECRETION

P. Groza, S. Cananău, E. Daneliuc and A. Bordeianu

Institute of Normal and Pathological Physiology, Bucarest, Rumania

The effect of acceleration and hyperthermia on the antidiuretic hormone secretion (ADH) was investigated in rats both separately and simultaneously. The two conditions of stress elicited a rise in plasma ADH concentration in proportion to their intensity. Concomitant exposure to the two factors produced an additional effect. The parallel histochemical studies using methods for demonstrating RNA, proteins and the neurosecretory material in the supraoptic nucleus, showed the synthesis and depletion of the hormone content in correlation with the plasma concentration of ADH.

1. Observations

The effect of acceleration and continuous hypergravity on organisms has been studied because of its practical interest in aviation and space flight. Thus, such symptoms as blackout and unconsciousness [10, 13] induced by stagnant hypoxia in aircraft pilots are well known. There have also been studies on the reactions of the body to hyperthermia. In both there are changes in blood distribution which result (besides other adaptative reactions) in increased secretion of ADH.

Acceleration induces a shift of circulating blood in the direction of the inertial force. Thus, with positive ($+G_z$) acceleration, blood is forced caudally into the lower abdomen and hind limbs, with emptying of the vascular bed of the cephalic, thoracic and upper abdominal regions [2]. Transverse hypergravitational stress influences blood flow to a lesser extent.

Brain hypoxia and the decrease in systemic blood pressure in the main reflexogenic areas give rise to a pressor response consisting of tachycardia and vasoconstriction, particularly splanchnic. A hormonal response also occurs which increases blood volume. This is mediated by increased secretion of ADH and aldosterone. The former causes increased reabsorption of renal water, whereas the latter acts by producing sodium retention. This is supposedly supplemented by renin oversecretion, resulting in an excess of circulating angiotensine. Initially, an increased ADH and aldosterone secretion is probably produced by the decrease in atrial stimulus of volume receptors and of carotid sinus and aorta baroreceptors [3]. Aldosterone is also stimulated by the decrease in the Na:K ratio (caused by an intracellular shift of Na and a K cellular outflow [2, 7], by angiotensine, as well as by ACTH). The two hormones contribute to the restoration of haemodynamic equilibrium by increasing the circulating blood volume.

The occurrence of hypergravitation-induced ADH oversecretion is shown by increase of ADH in circulating blood [2, 7], by reduced diuresis [17] and by histological alternations seen in the supraoptic nucleus and neurohypophysis [16]. The antidiuretic effect of increased gravity may be caused also by changes in renal circulation [4, 6].

The body reacts to hyperthermia by tachycardia, peripheral vasodilatation and increased blood volume (initially produced by mobilization of blood depots). The hypervolemic effect is further maintained by ADH and aldosterone over-secretion. The main role is played by ADH, since diuresis is not influenced by spirolactone which antagonizes aldosterone [1].

The occurrence of ADH secretion is shown by increased plasma [14] and urinary [12] levels, by reduced diuresis and by the fact that in rats and dogs [5] with hypothalamic lesions causing *diabetus insipidus* hyperthermia no longer results in hypodiuresis. Under these circumstances, an additional increase of blood volume is probably due to decreased gastric secretion, because this results from the administration of exogenous ADH [11]. A convergent effect on the hypervolemia is also that of thirst, enhanced by hyperthermia via the hypothalamo-preoptic region, which is also the site of osmoreceptors initiating ADH secretion. In hyperthermia, however, this mechanism is less involved; ADH secretion is primarily stimulated through the volume receptors in the central zone of the cardiovascular system, and presumably by stimulation of the peripheral thermoreceptors through their connections with the hypothalamus. A more generalized hormonal response may also occur, similar to a non-specific response to stress [8].

Experimental Material and Methods

Experiments were carried out on Wistar rats of both sexes, weighing 180 to 200 g. The animals were maintained at room temperature (23 °C) and at 37—40 °C and 42 °C, the animals being centrifuged for 15, 20, 30 and 60 minutes at $+6G_z$. They were divided into two series: in one series the blood ADH concentration was studied, in the second series the histochemistry of the supraoptic nucleus was studied. The rats were decapitated immediately after exposure to acceleration and hyperthermia, and blood samples and brain collected.

The antidiuretic hormone was assayed by the biological Dicker method [9], plasma AHD levels being expressed as microunits per millilitre.

The supraoptic nucleus was stained with the Brachet method for ribonucleic acid [18], the Danielli method for total proteins [19, p. 1016] and the Gomori method for the neurosecretory substance [19, p. 1444].

2. Results

2.1. Effect of Acceleration and Hyperthermia on Plasma ADH Concentration

The effect of acceleration and hyperthermia on plasma ADH concentration was studied by exposure to the two stressful agents either separately or concomitantly (Fig. 1). The results from rats at the temperature of 23° were taken as normal values.

Plasma ADH concentration depends on the amount of hyperthermia, the most intense response being obtained at 42°. Likewise, plasma ADH levels increase with increasing duration of hypergravity. When both factors are applied together the ADH secretion is still higher (Fig. 2). This stress is near the limits of tolerance, as it causes a 20% death rate.

N	1	2	3	4	5	6	7	8	9	10	11
N_2	11	7	8	11	12	10	9	8	9	9	12
0'	▨				▨	▨					
15'		▨					▨			▨	
20'								▨			
30'			▨								▨
60'				▨					▨		
23°	▨	▨	▨	▨							
37°					▨						
40°						▨ 42°				▨	▨
HAD	3,38	7,7	11,8	14,3	12,75	34,5	15,4	16,25	22	23,77	36,83
VSM ±	±0,26	±1,73	±3,07	±1,73	±2,7	±4,47	±2,23	±1,31	±1,90	±3,34	±5,47
P		< 0,1	< 0,01 < 0,05	< 0,001	< 0,005 < 0,001	< 0,001	< 0,001	< 0,001	< 0,001	< 0,001	< 0,001

Fig. 1. Schedule of investigations. Hatched colums designate the nature of experimental stress (time of hypergravitational stress and temperature of the ambient milieu). N = reference number of the group. N_2 = number of rats in the group.

2.2. Histochemical Changes in the Supraoptic Nucleus

Exposure for 30 minutes to acceleration at 23° activated RNA and protein synthesis, and produced an accumulation of neurosecretory material (Fig. 3a). After exposure for 30 min at 40° (without hypergravity) there was a tendency towards cytoplasmic RNA depletion, with only mild decrease in protein content. Also, as demonstrated by the Gomori reaction, there was a diminution of the neurosecretory material in the cell (Fig. 3b), but it was still present in the axis cylinders at the origin of the hypothalamo-hypophysial tract.

Hypergravitation for 30 min at 37 °C induced a moderate decrease in cytoplasmic RNA and proteins, and a depletion of the neurosecretory material (Fig. 3c). Hypergravitation for the same time at 40° causes a significant fall in cytoplasmic RNA, protein content and neurosecretory material. There was a substantial depletion in the cells (Fig. 3d).

In hypergravitation, ADH secretion is stimulated by the shift of blood volume and the drop in systemic blood pressure, triggering pressor reactions by means of the reflexogenic zones (endocardoaortic, sinocarotid, atria). The pressor reaction is further supported by CO_2 and H^+ accumulation as well by hypoxia and a part is also played by pulmonary athelectasia and the alveolar air-blood barrier changes [15].

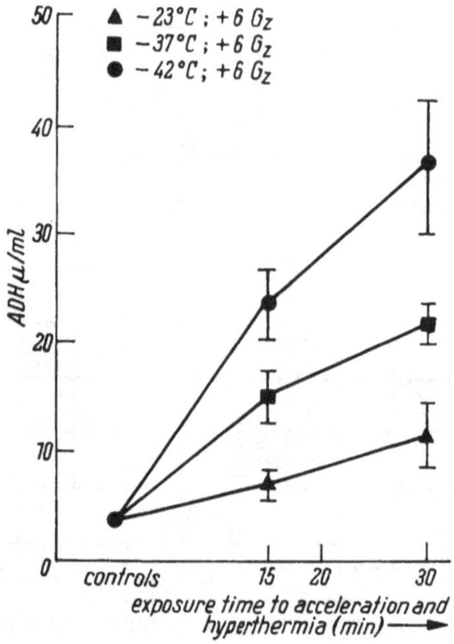

Fig. 2. Plasma ADH variations under the combined action of acceleration and hyperthermia.

References

[1] N. ABRAMSON and TH. E. PIEMME, Aerospace Med. **38**, 239 (1967).
[2] GR. BENETATO, S. CANANĂU and E. DANELIUC, Rev. roum. physiol. **7**, 15 (1970).
[3] G. C. BOND and J. W. TRANK, Am. J. Physiol. **222**, 595 (1972).
[4] I. P. BONVALLET, P. BERNCSÁTH and C. BOUFIGNAC, Am. J. Physiol. **222**, 2 (1972).
[5] M. BONVALLET and P. DELL, Arch. Intern. Physiol. **54**, 273 (1946).
[6] L. A. BRICKER, A. I. WAYNE, C. R. DAVIES and R. A. DOTTORO, Aerospace Med. **38**, 35 (1967).
[7] S. CANANĂU, Theses, I. M. F. Bucureşti 1971.
[8] K. I. COLLINS and I. S. WEINER, Physiol. Rev. **48**, 785 (1968).

Fig. 3. Histochemical changes in supraoptic nucleus under conditions of acceleration and hyperthermia.

1, Brachet method with methyl green pyronin for RNA.
2, Danielli reaction for proteins.
3, Gomori method with aldehyde fucsin for neurosecretory material.

N, supraoptic nucleus of normal rats.
a, acceleration ($+6G_z$) at 23 °C, 30 min.
b, hyperthermia 40°C for 30 min.
c, hyperthermia 37 °C + acceleration ($+6G_z$), 30 min.
d, hyperthermia 40°C + acceleration ($+6G_z$), 30 min.

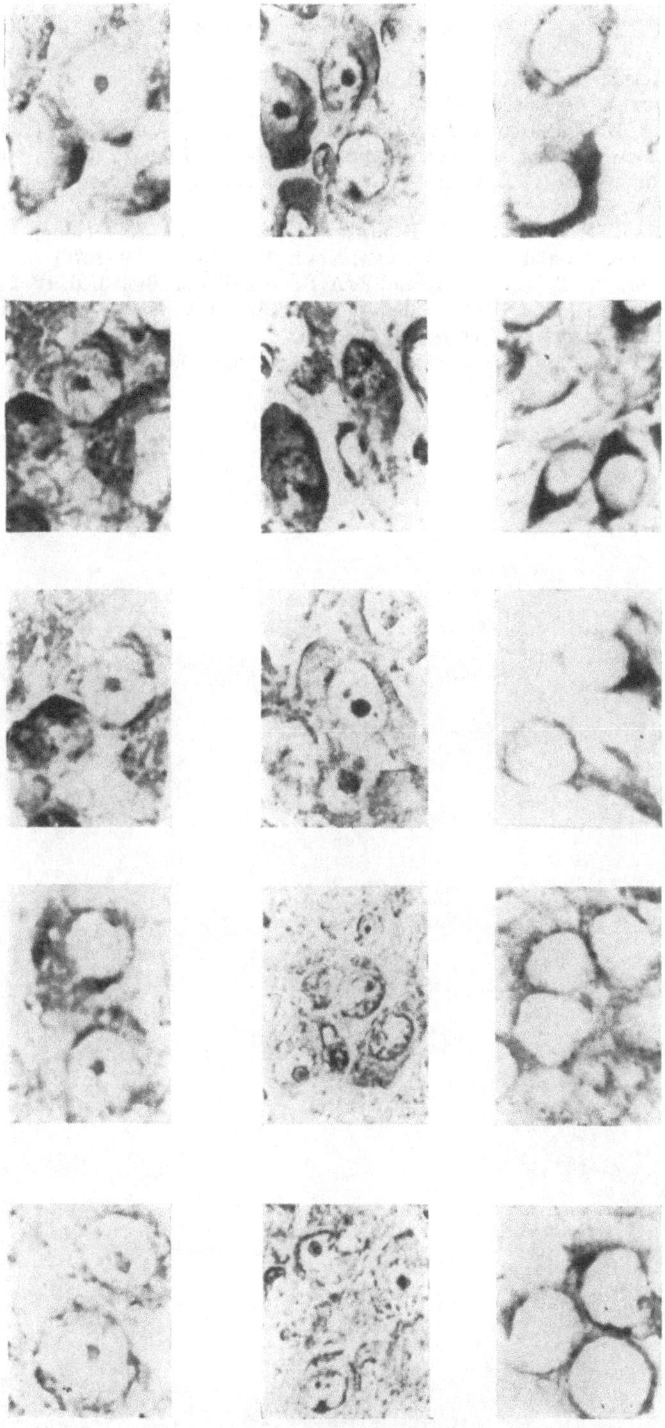
N)
a)
b)
c)
d)
1
2
3

[9] S. E. DICKER, J. Physiol. (L) **122**, 149 (1953).
[10] T. D. DUANE, Aerospace Med. **38**, 948 (1967).
[11] P. GROZA, M. CORNEANU and ST. IONESCU, Rev. roum. physiol. **3**, 121 (1966).
[12] K. HELLMAN and I. S. WEINER, J. Appl. Physiol. **6**, 194 (1953).
[13] L. A. LOHRBAUER, R. L. WILEY, Y. SHUBROOKS and M. MCCALLY, J. Appl. Physiol. **32**, 203 (1972).
[14] W. V. MACFARLANE and K. W. ROBINSON, J. Physiol. (L) **135**, 1 (1957).
[15] R. S. MOROZOWA and N. I. MARCHOVA, Kosm. Biol. Med. **3**, 29 (1970).
[16] A. S. PANKOVA, B. K. PODÎMOV and E. A. SAVINA, Kosm. Biol. Med. **12**, 21 (1970).
[17] TH. E. PIEMME, M. MCCALLY and A. S. HYDE, Aerospace Med. **37**, 1253 (1966).
[18] J. BRACHET, Quart. J. Micr. Sci. **94**, 1 (1953).
[19] A. G. E. PEARSE, in: Histochemistry, Vol. 2, Livingstone, Edinburgh and London 1972.

Life Sciences and Space Research XII — Akademie-Verlag, Berlin 1974

EFFECT OF DYNAMIC FACTORS OF SPACE FLIGHTS ON THE GREEN ALGA *Chlorella vulgaris*

E. V. MOSKVITIN and E. N. VAULINA

Institute of General Genetics, USSR Academy of Sciences, Moscow, USSR

The biological effects of vibrational and linear acceleration on the alga *Chlorella vulgaris* were studied. Periodic vibration in the frequency range of 4—4000 Hz with vibrational acceleration up to 16g did not affect the survival and mutability of *Chlorella* cells and did not modify the effects of acute γ-radiation. However, random vibration similar to that occurring during launch of spaceships, combined with linear acceleration increased the radiation damage to algae produced by acute γ-radiation at a dose of 10000 r. This effect is seen only in cells at the beginning of the G_1 stage, which precedes DNA synthesis.

1. Introduction

The unicellular green alga *Chlorella* is a convenient object for physiological and radiobiological investigations in space, and has often been used in space experiments. Our experiments on cultures of *Chlorella vulgaris* Beijer carried out in lunik Zond 8, the spaceship Soyuz 9 and the orbital station Salyut showed an increase in cell death (2.15%, 3.30% and 8.47% greater than the control), a decrease in the average number of autospores in the first post-flight sporulation, resulting in a decrease of the post-flight productivity of cultures. These results are in good agreement with the data obtained in earlier experiments [1—4]. The exposure of algae in the automatic station Zond 8 and orbital station Salyut caused an increase in the mutability of cultures. A similar effect was observed on the spaceship Soyuz 5 and lunik Zond 5 [2—4].

The action of dynamic factors—vibration and vibration combined with linear acceleration—is reported here, using the LARG 1 strain of *Chlorella vulgaris* Beijer.

2. Material and Methods

Physiologically active cultures of algae, nonsynchronous and synchronized by the change of light—dark conditions and by centrifusion, were used in the work. Cells, placed on the agarized Tamija medium, were exposed to the action of vibration [14]. Synchronous cultures included in the experiment were at the beginning of the G_1 stage of the cell cycle, in the middle of the G_1 stage or at the S-stage of DNA synthesis [5]. The choice of these stages was determined by the fact that they are the longest in the life cycle of *Chlorella* and dynamic factors at the launch of the spacechip are the most probable ones to influence cultures consisting of cells in these stages.

Vibration was carried out on an electrodynamic stand for 60 minutes in the frequency range of 4—4000 Hz and with vibrational acceleration up to 16 g. In order to reproduce the vibrational conditions during space flights as correctly as possible on the basis of data in the literature [6], a wideband random process was constructed and recorded with the maximum power-spectrum density (up to 0.1 g^2/Hz) of vibration in the frequency range of 300—800 Hz and the total average quadratic level of about 7—10g. The character of vibrational changes over time was similar to that at the launch of the spaceship. The recorded vibration was reproduced with the help of a tape recorder connected to the table of the vibrator. The simultaneous action of vibration and linear acceleration was carried out in a special centrifuge fixed on the vibrator. The vectors of vibrational and linear acceleration were mutually perpendicular; when combined with vibration the linear acceleration was 6—8g.

To study the modifying effect of dynamic factors on radiation damage, the *Chlorella* cells were exposed to γ-radiation at a dose of 10 kr at the rate of 570 r/min. The interval between the radiation and vibration was no more than 5 minutes.

The following tests were used to estimate the biological effects: the growth dynamics of the cultures, the average number of autospores in the first sporulation, disturbances in formation of autospores, productivity, survival and mutability. A modified method using microcolonies [7] was employed in order to study the development and survival of cells. During estimation of mutability macrocolonies with changed phenotype were taken into account—dwarf colony, pigment variants, speckled and morphological mutation (Fig. 1).

3. Results and Discussion

In preliminary experiments on non-synchronous cultures not a single test revealed any statistically significant differences between cells of the control and those exposed to vibration (in the frequency range of 4—40, 40—400, 400 to 4000 Hz). A weak tendency towards decrease in survival and increase in mutability was observed in cultures exposed to periodic vibration alone in the frequency range 400—4000 Hz in cells exposed to pre- and post-vibrational γ-radiation. If this difference was due to vibration but not random causes, it would have been possible to increase it by employing synchronous cultures of different ages. However, experiments conducted on synchronous cultures did not confirm this effect of vibration. Data on survival of synchronous cultures are summarized in Table 1.

The absence of effects after exposure of *Chlorella* to strictly periodic vibration does not prove there is no influence of vibration of the kind found during space flights. Many experiments were conducted with tape recordings of random wideband vibration similar in its parameters to that existing at different stages of space flight. A 45-minute model of such vibration, as well as experiments with strictly periodic vibration, did not reveal any action on intact synchronous cultures of *Chlorella*.

However, the combined action of vibration and post-vibrational γ-radiation on cells at the G_1 phase caused a statistically significant decrease in survival and

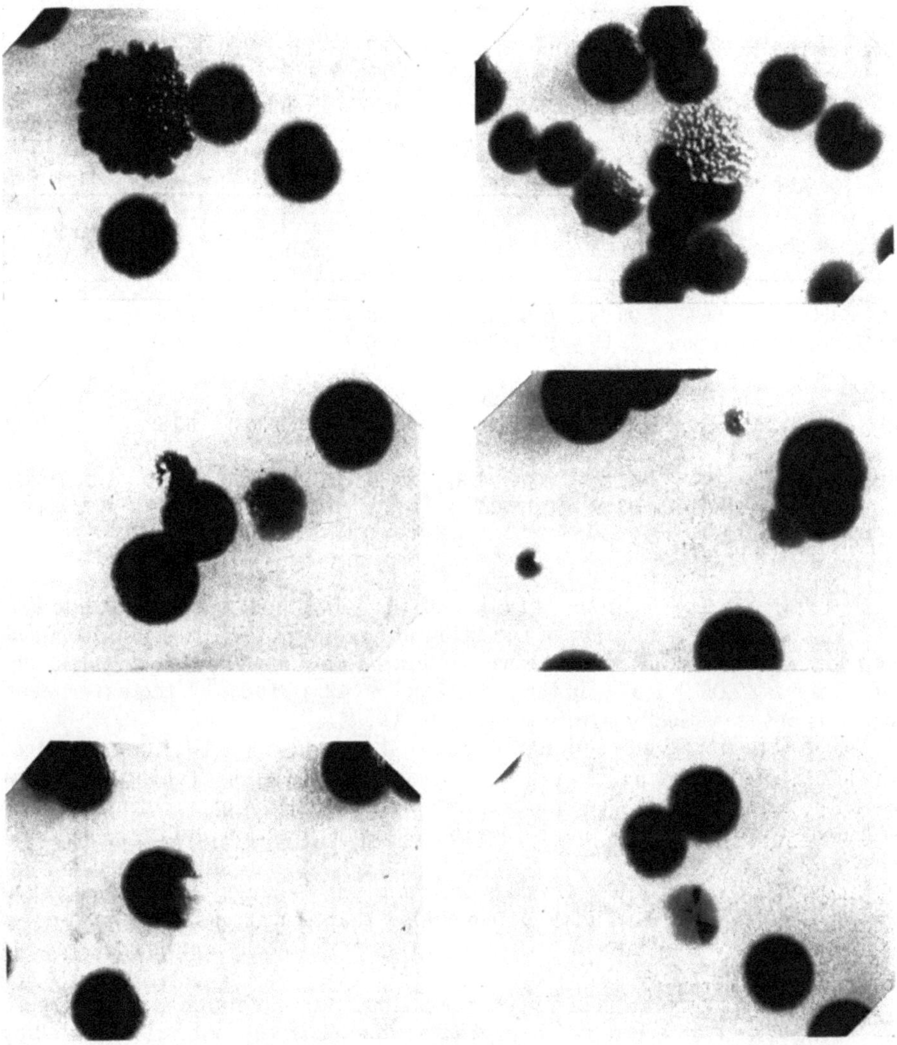

Fig. 1. Some types of phenotypically changed mutant colonies of *Chlorella vulgaris*. 1, 2, 4, morphological; 3, 5, 6, pigmental; 7, 10, light-coloured dwarf; 8, dark dwarf; 9, 11, sector pigmental.

an increase in the mutability of algal cultures. Fig. 2 shows the average survival of vibrated and non-vibrated cultures followed by γ-radiation based on three series of experiments consisting of 14 tests, each of which was repeated 3—4 times. All experiments show a decrease of the mean survival of cultures exposed to vibration. The average survival of control cultures was $79.87 \pm 0.41\%$ (9483 microcolonies examined) and that of vibrated cultures $73.17 \pm 0.46\%$ (9373 microcolonies examined). The mutability was respectively $6.31 \pm 0.17\%$ (19451 macrocolonies examined) and $7.46 \pm 0.20\%$ (15399 macrocolonies examined).

Table 1

The Survival of Synchronous Cultures of *Chlorella vulgaris* after Vibration in the frequency range 400—4000 Hz

The number of microcolonies counted is given in brackets

		Survival (%)		
		Without radiation	With radiation before vibration	With radiation following vibration
Beginning of G_1 stage	Control	93.08 ± 0.53 (2309)	65.86 ± 0.76 (3855)	64.27 ± 0.78 (3780)
	Vibration	92.69 ± 0.53 (2506)	65.40 ± 0.77 (3766)	64.48 ± 0.76 (3961)
		$t = 0.52$	$t = 0.43$	$t = 0.19$
Middle of G_1 stage	Control	92.30 ± 0.58 (2113)	53.65 ± 0.76 (4269)	56.65 ± 0.76 (4240)
	Vibration	92.71 ± 0.57 (2059)	52.74 ± 0.76 (4358)	55.82 ± 0.78 (3954)
		$t = 0.51$	$t = 0.84$	$t = 0.76$
S stage	Control	93.91 ± 0.46 (2678)	57.19 ± 0.75 (4372)	63.01 ± 0.74 (4352)
	Vibration	93.32 ± 0.49 (2529)	57.55 ± 0.74 (4465)	63.59 ± 0.75 (4119)
		$t = 0.88$	$t = 0.34$	$t = 0.55$

However, a decrease in the duration of the vibrational action to 9 minutes, i.e. the approximate duration of the firing of engines of the spaceship during its launch, reduced the difference between vibrated and non-vibrated cultures. The survival was 87.03 ± 0.47% in the control and 86.42 ± 0.48% in the experiment, which is not statistically significant ($t = 0.94$).

The combined 9-minute action of vibrational and linear accelerations reinforced the lethal and mutagenic effect of the subsequent γ-radiation (Tables 2, 3). However, the effect of these dynamic factors on non-irradiated algal cultures and on cultures exposed to γ-radiation before the action of vibration and linear acceleration was not tested.

Linear acceleration up to 15000 g does not cause changes in the survival of *Chlorella* [8]. Therefore it may be concluded that the acceleration we used of 6—8 g cannot by itself influence algal cells. Its significance is limited to increasing the vibrational effect.

Modification by vibration of γ-ray damage similar to that obtained in our experiments has been observed in *Drosophila melanogaster* [9]. Vibration preceding

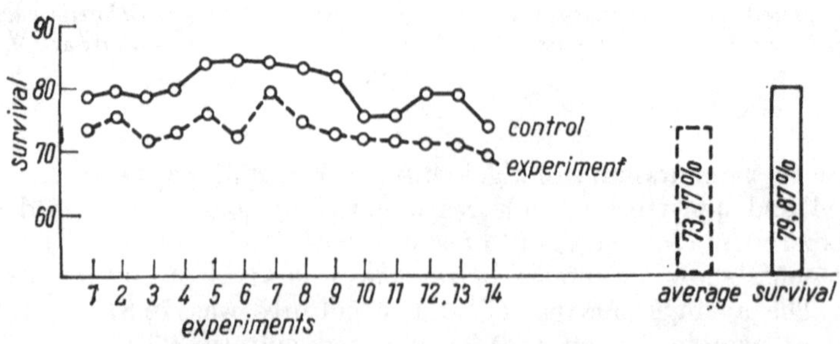

Fig. 2. The survival of *Chlorella* exposed to vibration and γ-radiation at the G_1 stage. —— control, --- experiment.

Table 2

The Combined Action of Vibration and Acceleration on the Survival of *Chlorella*

		Survival (%)		
		Without radiation	With radiation before vibration	With radiation following vibration
Beginning of G_1 stage	Control	98.99 ± 0.17 (3296)	88.66 ± 0.59 (2905)	91.06 ± 0.48 (5362)
	Experiment	99.28 ± 0.14 (2906)	88.53 ± 0.66 (2302)	86.59 ± 0.69 (2410)
		$t = 1.31$	$t = 0.15$	$t = 5.3$
Middle of G_1 stage	Control	99.69 ± 0.10 (2600)	82.24 ± 0.78 (2416)	83.11 ± 0.71 (2742)
	Experiment	99.49 ± 0.14 (2600)	81.75 ± 0.78 (2478)	82.50 ± 0.69 (3001)
		$t = 1.15$	$t = 0.63$	$t = 0.62$
S stage	Control	99.35 ± 0.17 (2434)	87.87 ± 0.71 (2118)	85.90 ± 0.62 (3131)
	Experiment	99.16 ± 0.17 (2500)	87.78 ± 0.60 (2996)	85.71 ± 0.79 (1930)
		$t = 1.12$	$t = 0.10$	$t = 0.20$

Table 3

The Combined Action of Vibration and Acceleration on the Mutability of *Chlorella*

		Survival (%)		
		Without radiation	With radiation before vibration	With radiation following vibration
Beginning of G_1 stage	Control	0.63 ± 0.10 (7357)	4.84 ± 0.24 (7630)	4.16 ± 0.20 (10003)
	Experiment	0.69 ± 0.10 (10311)	4.81 ± 0.28 (5928)	5.28 ± 0.30 (5772)
		$t = 0.60$	$t = 0.14$	$t = 3.7$
Middle of G_1 stage	Control	0.88 ± 0.10 (6795)	5.75 ± 0.30 (6211)	6.02 ± 0.28 (6810)
	Experiment	0.93 ± 0.10 (7494)	5.93 ± 0.30 (6241)	6.25 ± 0.28 (7308)
		$t = 0.50$	$t = 0.60$	$t = 0.82$
S stage	Control	0.62 ± 0.10 (10257)	5.83 ± 0.30 (5982)	5.79 ± 0.33 (4874)
	Experiment	0.58 ± 0.10 (7187)	5.69 ± 0.32 (5605)	5.71 ± 0.28 (6903)
		$t = 0.40$	$t = 0.32$	$t = 0.19$

γ-radiation at the spermatid stage noticeably increased the mutagenic action of γ-rays. N. I. Ribakov and co-workers [10, 11] observed a significant increase due to vibration of phage formation in *Escherichia coli*, induced by γ-radiation, though vibration itself did not induce phage formation.

Explanation of the mechanisms involved requires additional experiment. It may well be that the effect of vibration is determined by changes in certain enzyme systems of the cell. Suppression of the enzymic activity of myosine under the action of vibration has been shown [12, 13].

On the basis of our investigations we make the following conclusions:

(1) Strictly periodic vibration in the frequency range of 4—4000 Hz, or vibration of the type produced by working rocket engines, does not affect the survival and mutability of *Chlorella* cells and does not modify the effect of γ-radiation.

(2) Vibration and linear acceleration as found in space flight do not by themselves

cause essential physiological or genetic changes in *Chlorella vulgaris* cells. However, they are able to increase the subsequent radiation damage of algae, induced by a dose of 10000 r of γ-radiation. This effect is characteristic only of cells at the beginning of the G_1 stage of DNA pre-synthesis.

References

[1] E. N. VAULINA, I. D. ANIKEEVA and G. P. PARFYONOV, Kosm. Issled. **5**, 285 (1967).
[2] E. N. VAULINA et al., Kosm. Issled. **9**, 948 (1971).
[3] E. N. VAULINA et al., Life Sciences and Space Research IX, 105 (1971).
[4] I. D. ANIKEYEVA and E. N. VAULINA, Kosm. Issled. **9**, 946 (1971).
[5] I. D. ANIKEYEVA, Dissertation 1967.
[6] W. S. SHIPLEY and I. E. MACLAY, Astronautics and Acronautics **3**, 42 (1965).
[7] E. V. MOSKVITIN, in: "Matherialy YII Vsesojuznogo rabochego soveshchanija po voprosu krugovorota veshchestv v zamknutoji sisteme na osnove zshiznedejatelnosti nizshih organizmov", Naukova Dumka, Kiev 1972 (p. 5).
[8] Kratkij spravochnik po kosmicheskoj biologii i meditchine, *Chlorella*. Medicine, Moscow 1967.
[9] G. P. PARFYONOV, Kosm. Issled. **3**, 643 (1965).
[10] N. I. RYBAKOV and V. A. KOSLOV, Bulleten experimentalnoj biologii i meditchiny **6**, 64 (1966).
[11] N. N. ZSHUKOV-VEREZSHNIKOV et al., Problemy kosmicheskoj meditchiny. Materialy konferencii, Inst. Med. Biol. Problems, Moscow 1966 (p. 172).
[12] S. N. ROMANOV and R. A. ARTCHISHEVSKAJA, Biophysika **13**, 1054 (1968).
[13] S. N. ROMANOV, R. A. ARTCHISHEVSKAJA and V. A. MUUSEEV, Biophysika **14**, 233 (1969).
[14] H. TAMIJA et al., in: Algal Culture from Laboratory to Pilot Plant, Washington 1953 (p. 204).

Life Sciences and Space Research XII — Akademie-Verlag, Berlin 1974

DIGESTIVE AND RESORPTIVE FUNCTION OF THE SMALL INTESTINE IN STRESSFUL SITUATION

K. V. Smirnov and A. M. Ugolev

USSR

The present study examines the effects of severe stress on digestive and resorptive functions of the small intestine. Transverse 20-minute acceleration ($+10G_x$) results in an increase of invertase activity, particularly in distal parts of the small intestine. Although the activity of glycyl-l-leucine dipeptidase was changed, the fluctuations were less pronounced than those in invertase activity. Acceleration also produces rise in glucose accumulated in the intestinal mucosa and intensification of active carbohydrate transport. The displacement of the proximodistal gradient of invertase activity and carbohydrate resorption was significant. Dipeptidase and invertase activities in intestinal sacs and in tissue homogenates were depressed following exposure to 15-, 30- and 60-day hypokinetics. In distal parts the invertase activity was increased, and the resorptive function was enhanced.

Following exposure to unusual gaseous atmospheres (hypoxic, hypercapnic and hyperoxic) there was an increase in active glucose transport over the entire length of the small intestine.

Heat produces a decrease in invertase activity both in intact cells and homogenates. There were no pronounced changes in dipeptidase activity. Transport processes exhibited some variations.

1. Method

It is known that the alimentary tract is one of the most labile body systems which reacts sharply to physical factors and homeostatic changes, showing significant and continuous structural shift and functional alterations.

Such stress factors as acceleration and weightlessness affect the functional state of the small intestine, but the mechanism is still not clearly understood. Acceleration effects on the alimentary tract have been studied by some Soviet authors [1—6], but the effects of weightlessness have not been studied by either Soviet or foreign authors. Disorders of the digestive and resorptive functions of the small intestine in animals under changed environmental conditions have been observed by some workers [7—9].

We have investigated the stress effects of transverse acceleration, hypokinetics, altered gaseous atmosphere and heat on membrane digestion and intestinal resorption in rats. Invertase and glycyl-l-leucine dipeptidase activity was examined in isolated everted 2-cm-long sacs prepared from the small intestine, and in tissue homogenates of intestine. The former is indicative of the enzymatic activity on a surface of the intestinal epithelium, the latter shows the total amount of enzymes synthesized by the intestinal cells. Carbohydrate absorption from the small intestine was estimated by two methods that complemented each other. The method using everted intestinal sacs developed by Wilson and Wiseman [10] and modified by Ugolev and Loginov was employed. The advantage of this method is that it measures the activity of intact cells of the intestinal epithelium. It is essential also that digestive and absorptive processes are studied under the same conditions; this allows one to obtain data showing the interaction between the processes. The amount of glucose transferred from "mucosal" solution through the intestinal wall to "serosal" solution was determined in everted sacs of the small intestine, which were incubated in 0.2%-glucose at 37 °C under good aeration for 30, 60 and 90 min.

To study the initial absorptive processes (accumulation in the mucosa) we also employed a new method—"accumulating preparation of the intestinal mucosa" (APIM). This method has been developed at Ugolev's laboratory [11]. The amount of sugar absorbed was determined by the Nelson method [12] using Ugolev's [13] modification. The method is based on oxidation of reducing sugars by a copper reagent with subsequent reduction of arsenic-molybdenic reagent in the presence of the cuprous oxide formed. The work was carried out on albino male rats weighing 180—200 g. The animals were given no food for 18—20 hours before the experiment. The data obtained were evaluated statistically.

2. Effects of Acceleration

Rats were exposed to transverse acceleration for 20 minutes at $+10G_x$. The animals were studied 3, 6, 24, 48 and 72 hours after the exposure. The most pronounced changes in invertase activity on the surface of the intestinal cells appear to be found in the distal portions of the small intestine. The invertase activity was raised by 100—170%. The differences from the controls were statistically significant ($P < 0.001$). At the same time the invertase activity in the middle and proximal portions was increased by only 15—40%. The homogenates showed no such sharp difference in activity anywhere over the length of the small intestine. Invertase activity showed a sharp reduction only in the initial phase (3 and 6 hours) when its activity in the homogenates was decreased by 40—50% ($P < 0.01$). The invertase activity in the homogenates rose gradually and after 48 hours attained normal values. Thereafter there were no significant changes in invertase activity in the homogenates.

The study of glycyl-l-leucine-dipeptidase activity in the intestinal sacs and homogenates indicated that there were some changes, both on the surface of the cells and in the homogenates. The observed changes were specific to distal, middle and proximal portions. After 24 hours the enzyme activity reached a maximum level (an increase of 50—70%) and then decreased gradually. After 72 hours there were no significant changes in dipeptidase activity on the cell surface in the middle and distal portions of the intestine, compared with control values. However, in the proximal portions it was increased by 30%. The amount of the enzyme in homogenates of all portions was 30—40% above the control values.

In rate exposed to transverse acceleration there was an increase in glucose absorption. It was increased at 3 hours, and particularly at 6 hours. Thus, in control animals during three incubations an average of 2718 mg % of glucose was absorbed over the entire intestine; but in animals observed 3 hours after exposure to acceleration it was increased to 3677 mg %, and 6 hours after, it was

3894 mg %. After 24 hours active transport was somewhat decreased. The difference was statistically significant only in the two most proximal segments in the first 60 minutes. After 48 and 72 hours glucose absorption showed no difference compared with the control. (mg % = mg per 100 mg of tissue or blood.)

The amount of glucose in the intestinal mucosa of rats exposed to transverse acceleration was raised. This was observed at all the experimental periods. The greatest amount of accumulated glucose was observed at 3 hours following exposure. After 6 hours the amount of glucose was decreased slightly compared with the control. At 24 hours there was a moderate rise, but at 48 and 72 hours the glucose content in the intestinal mucosa of the test animals fell again, although it was still high compared with the control.

The data on glucose transport for the segments of the small intestine showed that the greatest glucose transport occurred in the 3rd and 4th segments (the entire small intestine was divided into six equal segments). In the control animals the 2nd segment was the most active. Thus, the test animals showed some displacement of the proximodistal gradient.

3. Effects of Hypokinesis

Hypokinesis was attained by means of special plexiglas cages that drastically reduced the animal's activity. Provision was made for feeding and drinking. The animals were examined at 15, 30 and 60 days.

The test animals showed a depressed activity of invertase and glycyl-l-leucine dipeptidase in the proximal segments, in both intestinal sacs and homogenates. Invertase activity was increased in distal segments.

The small intestine after hypokinesis for 15 days showed an increased amount of sugar in "serosal" liquid, which at the end of the first incubation (30 min) reached 1082 mg % (in the control, 455 mg %); at the end of the second incubation (60 min) it was 2058 mg % (in the control, 968 mg %); at the end of the 3rd incubation 1840 mg % (control, 792 mg %). If the total glucose transferred in all the segments (4980 mg %) is taken as 100%, the glucose transport in the 1st, 2nd and 3rd incubations is 21, 41 and 38% respectively.

Active glucose transport in the small intestine of rats exposed to 30 and 60 days of hypokinesis also showed an increase in sugar content with a linear dependence on the duration of hypokinesis. This regularity was most pronounced in the sacs incubated for 60 minutes. Active glucose accumulation followed the same relations, with the proximodistal gradient of glucose absorption at any of the hypokinetic periods displaced toward the distal end of the small intestine.

Immobilization stress was caused by fixing the animals for 17—24 hours in special cages. After stress for 1 day the invertase activity on the cell surface and in homogenates was markedly decreased in the test animals. After 72 hours there was a sharp increase of invertase activity both on the surface and in the homogenates, with the level of enzymatic activity elevated by 30% compared with the control. During immobilization the activity of glycyl-l-leucine dipeptidase fell significantly (by 77%) during the first day. Homogenate activity was decreased only by 18% during the first day, remaining at the same level for 72 hours. The most characteristic change in peptidases is their reduction on the surface of the cells. This reflects the translocation processes of the enzymes, whereas the smaller changes in activity in the homogenates reflects enzymatic synthesis.

4. Effects of Various Gaseous Atmospheres

Assessment of the active glucose transport in the small intestine was carried out on rats during four sets of experiments following exposure to: (i) normal gaseous atmosphere, (ii) hypoxic condition (ascending to an altitude of 12000 m at a rate of 2 m s^{-1}, exposure time 15 min), (iii) hypercapnic condition (CO_2 20%, O_2 21%, exposure time 12 hours), (iv) hyperoxic condition (1 hour exposure in a chamber with a 100% oxygen atmosphere pressurized to 4 atm). The chambers were maintained at a constant temperature of 20 ± 0.5 °C with a relative humidity of 70—90%.

Following exposure to various gaseous atmospheres there was an increase in the active glucose transport in all the segments. As a result the glucose concentration in "serosal" liquid was raised. In the first 30 minutes the greatest glucose transport in all the segments was seen after acute hypoxia (1664 mg %). At the same time under normal conditions the total amount of glucose transferred was 654 mg %, whereas under hypercapnic and hyperoxic conditions it was 1424 mg % and 1281 mg % respectively.

Comparison of the total amount of glucose which was transferred from "mucosal" to "serosal" solution due to active transport showed the greatest value in hypoxic animals (5926 mg %). It was 2.1 times as much as in normal controls (2777 mg %). Following hyperoxic and hypercapnic exposures it was 5254 mg % and 4760 mg % respectively.

Glucose was absorbed over the whole length of the small intestine. With all the stressful atmospheres transport activity was increased: 1.09—1.60-fold in the 1st and 2nd segments, 2—3-fold in the 3rd, 4th and 5th segments, 3—5-fold in the 6th segment (in comparison with the control). One sees that, in spite of a markedly increased activity in the distal intestinal segments, the glucose transport here was lower than in the proximal and middle segments.

5. Effects of Heat

Invertase and dipeptidase activity on the surface of the intact cells and in the homogenates was studied in albino rats in 4, 24, 48, 72 hours, and the following 6 days after a heat stress (2 hour exposure to a temperature of 39 °C). There were no marked changes in dipeptidase activity of the homogenates. The invertase activity was decreased in both intact cells and homogenates. There were also definite changes in transport processes.

When one compares the intensities between synthesis and translocation (referring to enzymes involved in membrane digestion) in animals exposed to single or repeated stressful effects, there is a big difference. Thus, the invertase activity was depressed more pronouncedly in adapted animals.

The data obtained show there is a delicate response of the small intestine to stressful factors. The adaptive changes occur over the entire intestine. There is an intensification in glucose intake, which in turn results in stimulation of enzymatic and transport activities.

A dominant part in regulation of the systems which call into action surface enzymatic action and transport appears to be played both by direct substrate regulation and also by the fact that the alimentary tract is a part of an integrated organism, and thus is under control of its nervous and humoral regulatory systems.

References

[1] I. M. Khazen, Nauchnoe Soveshchanic po Fiziologii i Patologii Pishchevareniya, Moscow 1958 (p. 51).
[2] I. M. Khazen, Deyatel'nost Pishchevapitelhoi Sistemi i iye Regulyatsiya w Norme i w Patologii, Moscow **1961** (p. 101).
[3] I. M. Khazen, Kosm. Biol. i. Med. **1**, 13 (1967).
[4] P. M. Suvorov, Bull. Exp. Biol. i. Med. **49**, No. 6, 54 (1960).
[5] V. E. Potkin, Bull. Exp. Biol. i. Med. **61**, No. 3, 43 (1966).
[6] K. V. Smirnov et al., Kosm. Biol. i. Med. **4**, 29 (1968).
[7] N. Sh. Amirov, Physiol. Zh. SSSR, **53**, 705 (1967).
[8] A. M. Ugolev et al., Dokl. Akad. Nauk SSSR **188**, 489 (1969).
[9] O. I. Babkina and K. V. Smirnov, Bull. Exp. Biol. i. Med. **12**, 30 (1970).
[10] T. H. Wilson and G. Wiseman, J. Physiol. **121**, No. 2, 45 (1953).
[11] A. M. Ugolev, Physiol. Zh. SSSR **56**, 651 (1970).
[12] Nelson, J. Biol. Chem. **153**, 375 (1944).
[13] A. M. Ugolev, Pristenochnoe (Kontaktnoe) Pishchevarenie, Moscow, Leningrad (1963).

Life Sciences and Space Research XII — Akademie-Verlag, Berlin 1974

RESPIRATORY GAS EXCHANGE AS AN INDICATOR OF CHANGED RADIORESISTANCE IN MAMMALS

L. NOVÁK and J. MIŠUSTOVÁ

Institute of Biophysics, Czechoslovak Academy of Sciences, Brno, Czechoslovakia

The paper deals with the problem of detection of radio-protective effects during irradiation. The method used is based on the fact that the known effective pharmacological and chemical radioprotectives (aminoethylthiouronium bromide HBr (AET), cystamine, sodium fluoroacetate (FAc), 5-hydroxytryptamine (5-HT), 5-methoxytryptamine (5-MT) and others) interfere directly or indirectly with enzymatic steps in energy metabolism of the organism. In mammals they induce at the same time an increase of resistance against ionizing radiation and a decrease in the respiratory gas exchange expressed by a depressed oxygen consumption.

In experiments carried out on male mice of the H strain and male rats of the Wistar strain it was found that animals respiring a hypoxic mixture (5—8% O_2, STPD), have a lower oxygen consumption during the period of increased radioresistance similar to that seen during the effect of radioprotective drugs.

This finding is of practical importance as it enables one to evaluate quantitatively the effect of radioprotective measures (including the hypoxia) in individual experimental animals during the course of irradiation.

1. Introduction

One of the important problems of space radiobiology is to find ways to increase the resistance of organisms against ionizing radiation. As far as the practical application of various radio-protective measures is concerned one of the basic problems is to discover a suitable method of detection of the effect of a radio-protective measure applied to a given individual organism during irradiation.

Increase in the resistance of mammals to ionizing radiation is always associated with a number of fundamental but transient changes in many physiological, and especially metabolic, functions [1]. This is expected because some radio-protective substances (AET, cystamine, FAc) interfere with enzymatic steps in energy metabolism. Other substances, for example the indolylalkylamines 5-hydroxytryptamine (5-HT) and 5-methoxytryptamine (5-MT), are effective indirectly because they act as vasoconstrictors, and thus reduce the oxygen supply to the cytochrome system. This results in a general decrease in the respiratory exchange. Using a method for continuous registration of respiratory exchange [2—5] we have found that increased resistance of the organism against ionizing radiation is associated with the period when the radioprotective substance induces in the individual a marked decrease in the oxygen consumption, reducing this to 40—60% of the original quiescent value. This observation has been confirmed by other authors [6—8].

It is of interest that no marked decrease in pO_2 in the tissue of a protected organism can be observed after the application of a number of radio-protective substances which would be comparable with the values found during respiration in a hypoxic mixture [9, 10]. Scarcely any attention has been paid in the literature to the problem of measuring the intensity of respiratory exchange of a mammal exposed to the effect of a hypoxic mixture in order to increase its radioresistance.

2. Methods

The estimation of respiratory exchange was carried out on male mice (H strain) and male rats (Wistar strain) by means of an apparatus designed in our laboratory [11, 12] using the total response of an automatic electric analyser. This enables one to follow the course of respiratory exchange after intraperitoneal injection

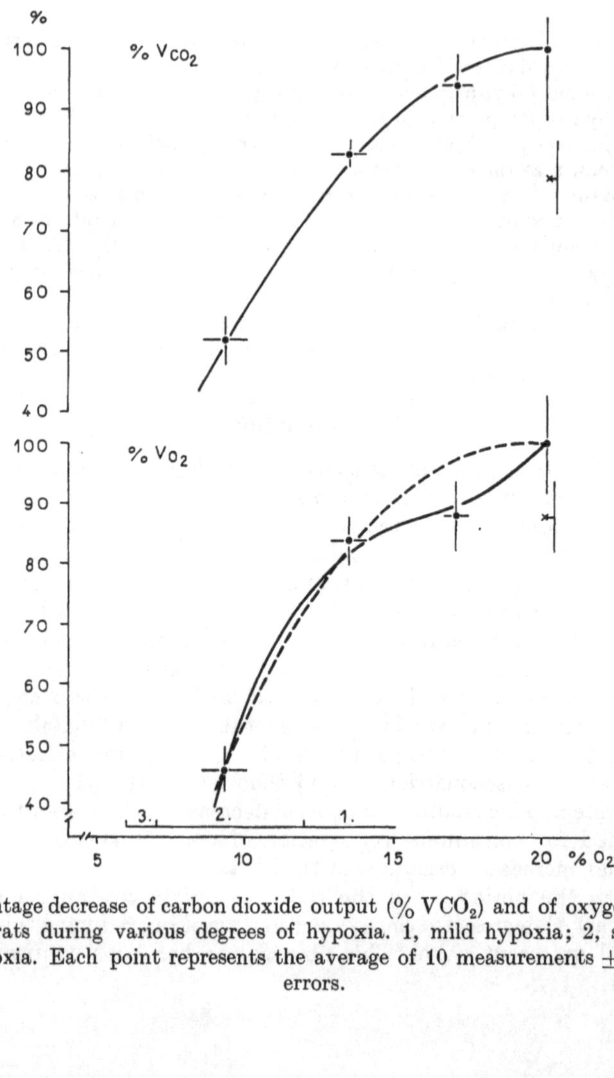

Fig. 1. Percentage decrease of carbon dioxide output (% VCO_2) and of oxygen consumption (% VO_2) of rats during various degrees of hypoxia. 1, mild hypoxia; 2, severe hypoxia; 3, acute hypoxia. Each point represents the average of 10 measurements ± their standard errors.

of various radio-protective substances even during the period of hypoxia induced in the metabolic chamber.

The radio-protective effectiveness of these measures was tested separately on mice by means of a single exposure to X-rays (800 R, 180 kV, 0.5 mm Al, 0.5 mm Cu, 50 R min^{-1}). This dose in our experimental conditions killed all control animals during a 30-day period. Experimental animals were irradiated at various intervals after the injection of radio-protective substance, 20—40 individuals for each point.

3. Results

The results of these observations are presented in Fig. 1. It is seen from the curves that in the region of hypoxia necessary to the increase in resistance against ionizing radiation—if the oxygen content in the inspired air is below 10% (STPD)—the oxygen consumption and carbon dioxide output of the animals decreases by 40—60% of the initial value observed at rest before the period of hypoxia. The decrease in the oxygen consumption occurs despite an increase in frequency of respiration, which during hypoxia is increased 100—150% above its initial value.

Fig. 2 shows that radio-protective hypoxia reduces the oxygen consumption, to the same extent as it is depressed by radio-protective doses of the drugs. Under these conditions all the radio-protective measures shown in Fig. 2 protected 90—100% of irradiated animals in time intervals marked by heavy lines, or

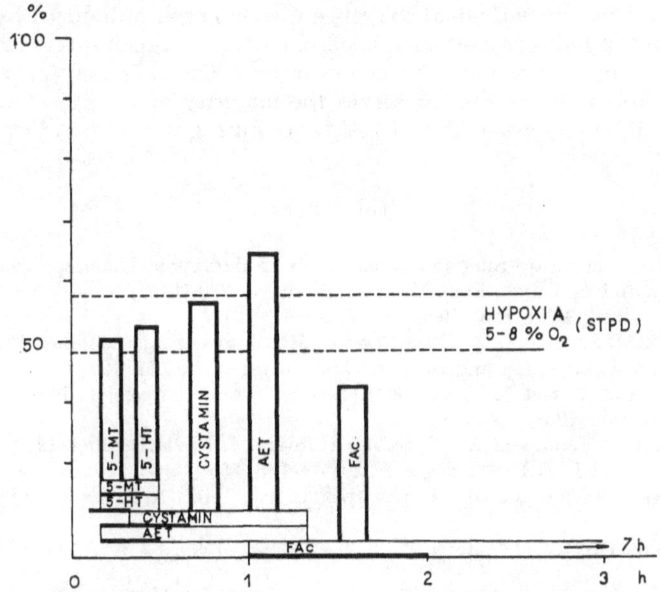

Fig. 2. Percentage decrease in the oxygen consumption of mice after intraperitoneal injection of 5-methoxytryptamine (5-MT) (25 mg kg^{-1}); 5-hydroxytryptamine (5-HT) (50 mg kg^{-1}); cystamine (200 mg kg^{-1}); aminoethylthiouronium bromide·HBr (AET) (200 mg kg^{-1}) and hypoxia (5—8% O_2; STPD) plotted against after injection of the drugs. For details see the text.

50–80% of animals in time intervals indicated by longitudinal belts. Also the peak depression of oxygen consumption measured from the injection of the drug is shown by the columns.

4. Discussion

The results presented suggest that the use of hypoxic mixtures containing 5–8% of oxygen (STPD) induces in the experimental organism (mice and rats) a decrease in the respiratory exchange by 40–60%. This probably reflects the decreased oxygen tension (pO_2) in the tissues which is in the range of values that limit the volume of oxygen utilized by the cytochrome system. Such an assumption is supported by the findings of Van der Meer et al. [13] and Vacek and Sugahara [14] according to which respiration of a mixture with an oxygen content of 5–8% (STPD) decreases the tension of oxygen dissolved in the spleen tissue to 2–4 torr at which level, according to Lehninger, oxygen becomes a limiting factor for the activity of the cytochrome system.

This implies that such an extraordinarily effective form of radio-protective measure as hypoxia has a common denominator with other means of pharmacological protection in the decrease in metabolic processes utilizing oxygen in the protected organism. This phenomenon provides the possibility of controlling the radio-protective effect of hypoxic mixtures, as well as of the radio-protective drugs, in a given individual during irradiation by measuring the oxygen uptake and carbon dioxide output. From the practical point of view this is of especial importance under conditions of space flight.

There is a different individual tolerance to the length of induced hypoxia and to the amount of radio-protectors applied in individual organisms. The method of continuous measuring of the level of respiratory exchange may provide a very valuable means of study that preserves the integrity of the object studied and allows a sensitive regulation of radioresistance during the course of irradiation.

References

[1] Z. M. Bacq, Chemical Protection against Ionizing Radiation, Thomas, Springfield, 1965.
[2] L. Novák and A. Vacek, Folia biologica (Praha) 4, 209 (1958).
[3] L. Novák, Bull. Acad. Roy. Belg. Ser. V, **52**, 633 (1966).
[4] L. Novák, J. Šikulová, B. Hošek and J. Mišustová, Radiation Res. **40**, 430 (1969).
[5] J. Mišustová, L. Novák and B. Hošek, Strahlentherapie **141**, 126 (1970).
[6] E. F. Romantsev and A. V. Savitsch, Chemical Protection against Ionizing Radiation, Medgiz, Moscow 1958.
[7] E. H. Betz, P. Lelievre and V. Smoliar, Intern. J. Radiation Biol. **12**, 163 (1967).
[8] A. Locker and J. E. Pany, Z. ges. exp. Med. **138**, 331 (1964).
[9] E. Ya. Grayevsky and M. M. Konstantinova, Dokl. Akad. Nauk SSSR **133**, 969 (1960).
[10] E. Ya. Grayevsky and M. M. Konstantinova, Dokl. Akad. Nauk SSSR **140**, 705 (1961).
[11] B. Hošek, L. Novák and K. Štefka, Čs. fysiol. (Praha) **11**, 121 (1962).
[12] L. Novák, B. Hošek and M. Kukleta, Studia biophysica (Berlin) in press (1974).
[13] C. Van der Meer, D. W. Van Bekkum and J. A. Cohen, Proc. 2nd UN Conf. on Peaceful Uses of Atomic energy, **23**, 42 (1958).
[14] A. Vacek and T. Sugahara, Proc. Intern. Conf. Radiation Biology and Cancer, Kyoto 1966, Radiation Society of Japan 1967 (p. 77).

Life Sciences and Space Research XII — Akademie-Verlag, Berlin 1974

METABOLIC RESPONSES OF MONKEYS TO INCREASED GRAVITATIONAL FIELDS

A. H. SMITH, D. F. RAHLMANN, A. M. KODAMA and N. PACE

Department of Animal Physiology and Department of Physiology, University of California, Davis and Berkeley, USA

A depletion of body fat has been reported by several investigators for various species of homoiotherms exposed to chronic acceleration. Generally the effect is proportional to field strength, and over the size range examined (0.25—5 kg) it becomes more severe in larger animals. There is some evidence that this may be the result of increased secretion of a pituitary hormone, the fat mobilizing substance (FMS).

An exception to the defatting effect of chronic acceleration was observed in mature pig-tailed monkeys (*Macaca nemestrina*)—although they exhibited other changes generally found in homoiotherms exposed to hyperdynamic environments. It is considered that this unusual primate response to chronic acceleration may result from a different geometry of the central nervous system, and its effect upon gravity load distribution in the lower part of the brain.

Many physiological responses of homoiotherms to continued residence in hyperdynamic fields have been reported for a variety of species and by several investigators; these have been reviewed previously [1—3]. One of the more frequently observed changes is a visibly obvious reduction in body fat content, which has been encountered in mice [4, 5], hamsters [6, 7], rats [8—12] and chickens [13—15]. Where studies involved observations at several acceleration intensities, it appeared that this effect was proportional to field strength. Also, among species the degree of defatting for equivalent acceleration increases with increasing animal size. The loss of fat is associated with an increased energy metabolism (e.g. a greater maintenance feed requirement [16]), and changes in tissue enzyme activities which indicate shifts in intermediate metabolism, favoring enhanced carbohydrate utilization [9, 17—20]. These changes are not the result of an inability to secure feed, or of a biosynthetic limitation. When chronically accelerated chickens are subjected additionally to a 3-day fast, they may lose 300 grams of body mass. When feeding is recommenced this deficit is recovered as quickly in the centrifuging animals as in the gravity-controls [21], indicating that the reduced body mass, and the altered body composition are physiological phenomena. Some evidence, reviewed in [2], indicates that a fat mobilizing hormone is involved, which appears to be similar to the "FMS" (fat mobilizing substance) responsible for human lipodystrophy [22, 23].

In 1971, we undertook the chronic acceleration of mature male pig-tailed monkeys, which generally have body masses over 10 kg. It was anticipated that such large animals would be quite susceptible to chronic acceleration on the basis

Table 1

Acceleration Schedule

The days-duration of exposure to each field is indicated individually. The principal intensities were: $1G$, $1.5G$, $2.0G$ and $2.5G$, with the interim periods arranged to enhance physiological adaptation.

Acceleration Field, G	1.1	1.25	1.4	1.5	1.75	1.9	2.0	2.3	2.5	Total Days Exposure
Monkey Name and Colony No.					Days Exposure					
290, Lymoges	8	7	7	54	5	2	118	3	8	212
374, Green	8	6								14
391, Dorset	7	7	7	33	5	2	118	3	11	193
396, Lovel	7	7	7	33	5	2	118	3	32	214
409, Urswick	7	7	7	33	5	2	118	3	34	216

of the tolerance of smaller animals to the treatment; these tolerances are: mice, 30 g, $7G$ [5]; rats, 250 g, $5G$ [10] and chickens, 2 kg, $3G$ [13]. Consequently, the acceleration schedules arranged for these monkeys were rather cautious (Table 1). As with our other chronic acceleration experiments, the treatment involved brief daily interruptions, generally 10 minutes for feeding, watering and cage cleaning. At weekly intervals the treatment was interrupted for about an hour to permit more detailed examinations of the animals. However, since there appears to be a time-intensity summation of acceleration effects upon vegetative phenomena [24] these interruptions were not considered to have a significant influence upon the results obtained.

Four of the five chronically accelerated monkeys tolerated fields up to $2^1/_2 G$, close to the limit observed in much smaller animals. Also, except for "Green" (No. 374) the cause for terminating the treatment was mechanical (e.g. surface abrasion) rather than the chronic acceleration sickness observed previously [13]. Consequently, it appears that the chronic acceleration tolerance of primates may be greater than would be predicted from experience with smaller homoiotherms.

At monthly, or longer intervals, the chronic acceleration was suspended for several days, and the body fat content of the monkeys was estimated from body water measurements [25]; the results are summarized in Table 2. It is evident that the chronic acceleration-induced loss of body fat seen in other homoiotherms is not shared by monkeys: on the contrary there was a marked and progressive increase in body fat concentration during the period of centrifugation. Since this greater body fat content was not reversed even after 50–67 days residence at earth gravity (Table 2), it would appear to represent a developmental process, a physiological change in body composition rather than a direct result of the chronic acceleration. It appears that the process which causes the spectacular defatting in other chronically accelerated animals (presumably an excessive production of FMS) is either not present in monkeys or is not activated during their centrifugation.

Studies by others indicates that the FMS hormone is produced by the pituitary in mammals and by the hypothalamus in birds [26]. It is possible that the observed increased production of FMS during chronic acceleration may result from

Table 2

Body Composition of Chronically Accelerated Monkeys

The body masses, absolute and relative fat and lean body contents are summarized for the indicated period of treatment as the mean ± the standard deviation.

	Mean ± SD				
	Body mass (kg)	Body fat (kg)	(%)	Lean body mass (kg)	(%)
Pre-acceleration	12.23 ± 1.54	0.55 ± 0.18	4.10 ± 1.10	11.68 ± 1.49	95.9 ± 1.1
4 wks at 1.5 G	11.29 ± 1.38	0.57 ± 0.13	5.00 ± 0.84	10.72 ± 1.27	95.0 ± 0.8
4 wks at 2.0 G	11.11 ± 1.09	0.54 ± 0.16	4.80 ± 1.97	10.57 ± 0.99	95.2 ± 1.9
16 wks at 2.0 G	11.38 ± 0.71	0.69 ± 0.25	6.02 ± 2.10	10.69 ± 0.66	94.0 ± 2.2
8–34 d at 2.5 G	11.07 ± 0.97	1.04 ± 0.34	9.47 ± 2.77	10.03 ± 1.05	90.5 ± 2.8
Recovery, 50–67 d at earth gravity	11.78 ± 1.26	1.00 ± 0.34	8.35 ± 2.24	10.78 ± 1.06	91.6 ± 2.2

mechanical stimulus of the region of the brain in which it is produced—a situation somewhat like the release of renin by the kidney in response to excessive hemodynamic forces. This possibility is consistent with the negative buoyancy of the brain, which would be quantitatively proportional to the ambient acceleration

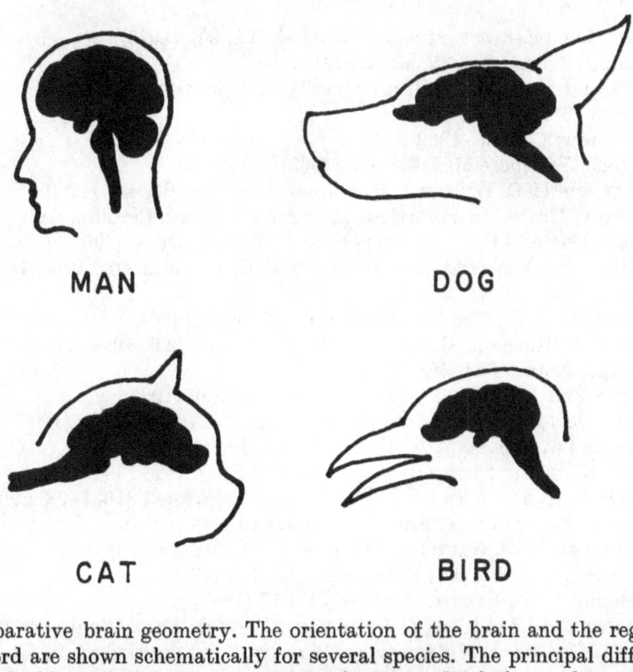

Fig. 1. Comparative brain geometry. The orientation of the brain and the region of entry of the spinal cord are shown schematically for several species. The principal difference is found in man, in which the cord enters approximately perpendicularly to the major axis of the brain, and more or less below the center of mass. The possible relationships of these differences in geometry upon the response to chronic acceleration are discussed in the text.

field, so that equal increments in field strength would lead to equal increments of brain loading and perhaps a rectilinear release of FMS. On this basis, the geometry of the brain and the influence of the point of entry of the spinal cord were considered as potential factors in the load distribution of the brain, and the likelihood of a hyperdynamic environment leading to a mechanical stimulus of the hypothalamic-pituitary region.

In Fig. 1, the geometric relationships of spinal cord and brain are shown, schematically, for several homoiotherms. It is obvious that for the quadrupedal mammals, and also for birds, the cord joins the brain posteriorally, and at an oblique angle. Such an arrangement would provide the cord with very little influence over any gravitationally induced loading of the brain. In man, and presumably in the primates generally, the entry of the cord into the brain is more central, and perhaps more or less under the center of mass. Such an arrangement could protect the hypothalamic region from mechanical stimulation, and perhaps account for the difference in metabolic responses of monkeys and small homoiotherms to chronic acceleration.

Acknowledgment

This study was supported under National Aeronautics and Space Administration grants NGR 05-004-008 and NGL 05-003-024.

References

[1] A. H. SMITH, Life Sciences and Space Research XI, 201 (1973).
[2] A. H. SMITH, J. Animal Sci. **35**, 635 (1972).
[3] A. H. SMITH and R. R. BURTON, in: Gravity and the Organism. University of Chicago Press, Chicago 1971.
[4] L. C. KEIL, Growth **33**, 83 (1969).
[5] C. C. WUNDER, Aerospace Med. **33**, 866 (1962).
[6] S. R. BRINEY and C. C. WUNDER, Proc. Iowa Acad. Sci. **67**, 495 (1960).
[7] P. C. CANONICA, Master's dissertation, University of South Carolina (1966).
[8] H. W. CASEY, Doctoral thesis, University of California, Davis (1965).
[9] D. D. FELLER, E. D. NEVILLE, J. OYAMA and E. G. AVERKIN, Proc. Soc. Exp. Biol. Med. **19**, 522 (1965).
[10] J. OYAMA and W. T. PLATT, Am. J. Physiol. **209**, 611 (1965).
[11] G. C. PITTS, L. S. BULL and J. OYAMA, Am. J. Physiol. **223**, 1044 (1972).
[12] F. L. D. STEEL, Nature **193**, 583 (1962).
[13] R. R. BURTON and A. H. SMITH, Aerospace Med. **36**, 39 (1965).
[14] J. W. EVANS, Doctoral dissertation, University of California, Davis (1968).
[15] A. H. SMITH and C. F. KELLY, N. Y. Acad. Sci. **110**, 410 (1963).
[16] A. H. SMITH, R. R. BURTON and C. F. KELLY, J. Nutr. **101**, 13 (1971).
[17] J. W. EVANS, A. H. SMITH and J. M. BODA, Am. J. Physiol. **216**, 1468 (1969).
[18] J. W. EVANS and J. M. BODA, Am. J. Physiol. **219**, 893 (1970).
[19] D. D. FELLER and E. D. NEVILLE, Am. J. Physiol. **208**, 892 (1965).
[20] B. C. DALIGCON and J. OYAMA, Physiologist **13**, 174 (1970).
[21] A. H. SMITH and R. R. BURTON, Growth **31**, 317 (1967).
[22] T. M. CHALMERS, in: Handbook of Physiology, Waverly Press, Baltimore 1965.
[23] B. SENIOR, in: Handbook of Physiology, Waverly Press, Baltimore 1965.
[24] R. R. BURTON and A. H. SMITH, Am. J. Physiol. **222**, 1505 (1972).
[25] A. M. KODAMA, J. Appl. Physiol. **29**, 260 (1970).
[26] I. NIR, M. K. DIMICK and S. LEPKOVSKY, Canad. J. Physiol. Pharmacol. **47**, 435 (1969).

Life Sciences and Space Research XII — Akademie-Verlag, Berlin 1974

OSSEOUS MALREPAIR IN CALCIUM-DEFICIENT STATES

J. R. Beljan

School of Medicine, University of California, Davis, Calif., USA

A primary body tissue involved in metabolic change under the influence of gravity is bone. Osseous integrity is obviously essential for normal musculoskeletal activity; this is of particular importance in long manned space flights where some evidence suggesting calcium loss from bone exists. Studies in our laboratory, utilizing an avian model, have been on the effects of chronic calcium deficiency (or "nutritional secondary hyperparathyroidism"), particularly on bone repair. Quantitative methods to determine both the rate and quality of osseous defects have been developed. A calcium-deficient synthetic diet (0.002% calcium by weight, but otherwise nutritionally complete) has been formulated to permit the induction of true calcium deficiency states in adult, skeletally mature, male domestic fowl. This report is concerned with the effects of long-term calcium depletion on bone repair. It has been classically taught that bone injury repair is always preferentially treated by the body, despite skeletal calcium depletion. For the first time, we have been able to demonstrate that this is incorrect; in the chronic calcium-deficient state, an osseous defect will *not* be preferentially calcified at the expense of an already-decalcifying skeleton.

Quantitative data demonstrate that healing patterns in bone vary in response to the duration of the calcium deficiency. In adult male fowl, calcium-deficient for ten months, bone repair is almost completely absent; in those calcium-deficient for six months, bone repair is both retarded and imperfect; repair is absent even when supplemental calcium is administered at the time of osseous injury induction and thereafter.

The conclusions to be drawn from this study are that (i) bone repair is adversely affected in the calcium-deficient state, (ii) the rate and quality of bone repair are directly related to the duration of the calcium-deficient state, and (iii) supplemental calcium administration in chronic calcium-deficient states initially retards, rather than enhances, bone repair. These conclusions have direct relevance to long-term manned space flight.

1. Introduction

Many studies have been done to explore the effects of low dietary calcium intake on the skeleton of both growing and adult animals [1—9]. It is generally accepted that the calcium pool represented by the skeleton will supply endogenous calcium if dietary calcium is inadequate. The present study was conducted to examine the effect of chronic decalcification on the repair of an experimental cortical bone defect. The decalcifying state is achieved by feeding a highly purified diet based on blood fibrin as a protein source [2]: the calcium content of this diet is 0.002% by weight. Studies in this laboratory have shown this level to be below the MDR (minimum daily requirement) for the rooster, resulting in skeletal calcium loss [7].

2. Measurement Method

The standardized experimental osseous defect used in this study and the methods developed to measure the rate of repair are described in [1, 2]. Basically, a $^3/_{32}$-inch circular defect is made in the lateral cortex of the avian metatarsus, at approximately midshaft. To permit later external location of the defect (osteotomy), two transfixing Kirschner wires are placed on either side of the osteotomy, 1 cm distally and 2 cm proximally. The Kirschner wires also serve as the points of attachment for a specially fabricated acrylic jig which permits accurate and reproducible placement of the osteotomy site and a specific control site in the path of a collimated monoenergetic photon beam [1].

The photon beam is emitted from a ^{125}I source originally adapted by Cameron et al. [10, 11] to serve as a method of photon densitometric mineral analysis. The Cameron method was modified by this laboratory to assess accurately the rate of bone repair quantitatively by a comparison of transmission rates through the osteotomy and a control site 1 cm proximal to the osteotomy [2]. Expressed by the exponential law of attenuation, these values would be:

$$I_a (\text{or } I_b) = I_0 e^{-\mu x}$$

where subscripts a, b, and 0 identify transmission rates of the osteotomy, normal bone and air respectively. A ratio of the attenuation values of the osteotomy and control sites,

$$\frac{\ln (I_0/I_a)}{\ln (I_0/I_b)}$$

constitutes the basis of the fracture repair ratio [1, 2]. Changes in this ratio (with the assumption that all contributing parameters remain constant except for callus formation in the osteotomy) is a function of bone mineral deposition in the osteotomy.

3. Experimental Methods

Two separate experiments were conducted with adult male Single Comb White Leghorn domestic fowl. In the first experiment eight animals were maintained on a low calcium purified diet [7] containing 0.002% calcium by weight for ten months prior to inducing the fracture; an additional four animals of the same age were used as controls and remained on commercial poultry rations. The second experiment consisted of nine animals maintained on low calcium rations for six months prior to wounding. At the time of wounding, the latter animals received supplemental dietary calcium (0.1%); four additional animals on commercial rations were used as controls. All animals were housed in individual layer cages and given food and demineralized water *ad libitum*. Photon-beam transmission data were collected by averaging six consecutive six-second counts daily for thirty days for each transmission rate (osteotomy, control site, and air). Attenuation values were calculated and normalized for each animal by setting the initial value of the osteotomy to 0.800 and the control site to 1.000. These values were chosen based on group means. Both black-and-white photographs and radiographs were taken of each metatarsus at the termination of the experiment to permit gross qualitative evaluations of the osteotomy. Radiographs were taken with a Faxitron® using Kodak Type R Industrial Film and an exposure time of fifteen minutes. High copy contrast film was used in the standard photograph.

The health of the animals was evaluated by monitoring body weight, general appearance, and by performing standard hematological analyses. In addition, bacterial cultures were performed on all osteotomies in the second experiment at the time of sacrifice to exclude the possibility of wound infection.

4. Results and Discussion

Results from the first experiment (low calcium rations, ten months) are given graphically in Figs. 1—3; all data points are group means. Fig. 1 describes the attenuation of the control site for both experimental and control animals. The data indicate that bone mineral is being lost in the calcium-deficient state; this is shown by the slightly negative slope of the low-calcium curve compared with the slightly positive slope of the control curve. The most dramatic effect of chronic calcium deficiency is the lack of normal osseous repair, as shown in Fig. 2. Representative photographs (Fig. 4) and radiographs (Fig. 5) confirmed this finding;

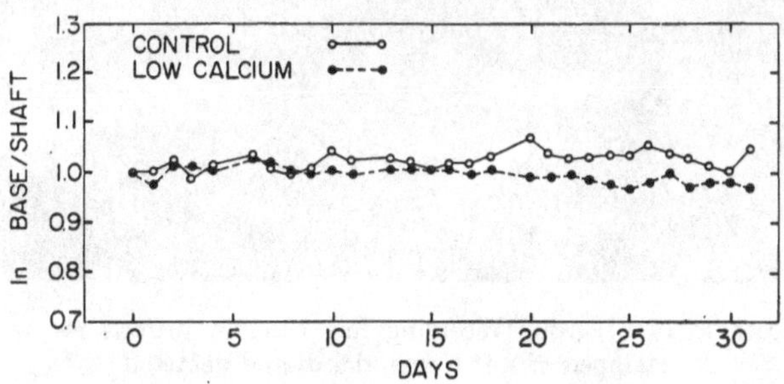

Fig. 1. Relative bone mineral mass of the control site 1 cm proximal to the osteotomy: control animals and experimental animals fed low calcium rations for ten months (mean of the standard deviations for each curve less than 0.0661).

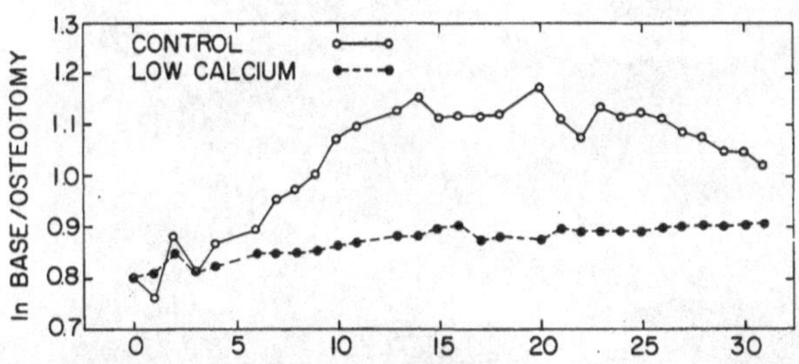

Fig. 2. Relative bone mineral mass in the healing osteotomy: control animals and experimental animals on ten months low calcium rations (means of the standard deviations less than 0.1653 for control curve and 0.0747 for the experimental curve).

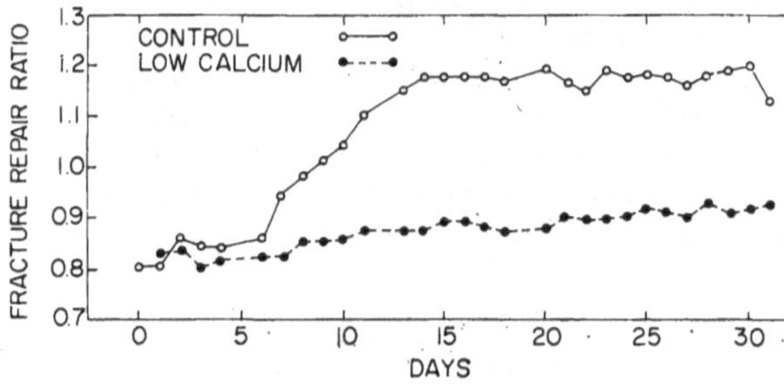

Fig. 3. Fracture repair ratio of control animals and experimental animals on ten months low calcium rations.

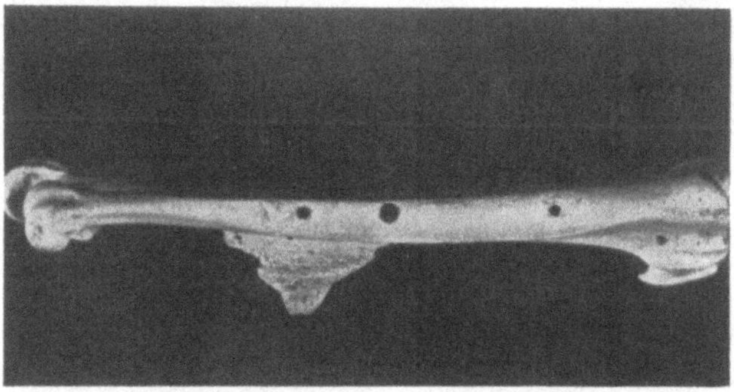

Fig. 4. Typical osteotomy of group receiving low calcium rations for ten months. Bone stripped of soft tissue, dried, and defatted.

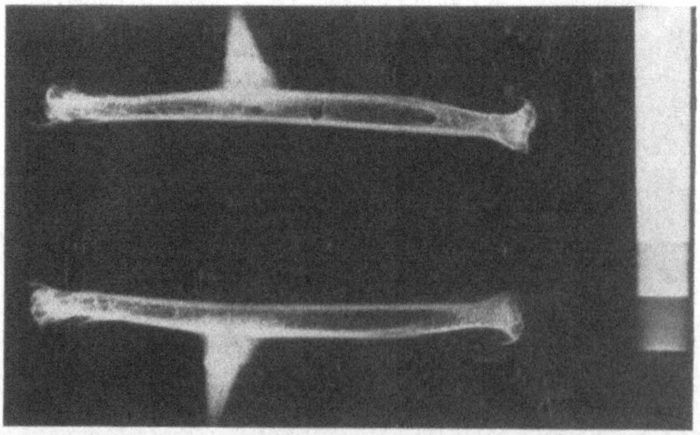

Fig. 5. Typical osteotomy of group receiving low calcium rations for ten months. (Control bone included for comparison.)

no signs of wound infection nor any other abnormality were observed at the time of sacrifice. The apparent conclusion is that the animal in the truly chronic calcium-deficient state will not preferentially calcify an osseous defect at the further expense of an already decalcifying skeleton.

Other explanations for the lack of repair must obviously also be considered. Some of these possibilities are: (i) the osseous defect was not of sufficient magnitude to demand precedence over total skeletal integrity; (ii) the onset and/or rate of repair was so drastically delayed that significant repair was not instituted during the four-week period; (iii) the calcium pools normally used for repair were either unavailable or totally depleted.

The second experiment of this series was designed to examine several of the above possibilities. If the duration of the state of decalcification influences the rate and magnitude of repair, intermediate healing patterns should be expected from animals placed on low calcium rations for shorter periods of time. Support for this hypothesis lies in the results of the original experiment [2], in which low calcium rations were given only two weeks prior to wounding; healing in that study occurred fairly normally, although the callus did not attain a density measurement comparable with control values, and resorption during the remodelling phase seemed to be enhanced. If, however, sufficient endogenous calcium pools were not available for bone repair, supplemental exogenous dietary calcium should ameliorate this situation. The results of the second experiment are given in Figs. 6–8. The response of the control site is the same as that observed in the first experiment; the unwounded bone shows a decalcifying trend in animals receiving low calcium rations as well as those being given supplemental calcium after six months of calcium deficiency. Repair of the defect, as indicated by osteotomy calcification (Fig. 7), did exhibit an intermediate response for those animals on the low calcium regimen. However, the repair curve of the group receiving supplemental calcium was identical with the curve of the low calcium group observed in the first experiment (Fig. 2), indicating minimal repair. These results are the reverse of what could have been anticipated, and every effort was made to assure their validity. Repair curves of all individual animals were examined and the group curves in Fig. 7 were found to reflect accurately the individual responses.

The fracture repair ratios of all groups (Fig. 8) are similar to the repair curves of the osteotomy (Fig. 7). The fracture repair ratios of the low calcium group and the control group are more similar than the curves of the osteotomy attenuations because of the diverging pattern of their respective control site curves; that is, values of the control fracture repair ratio will tend to be less than osteotomy attenuation values because of the increasing control site values. The opposite situation exists with the low calcium group where control site values are decreasing.

The results of this study indicate that the rate of fracture repair of the domestic fowl is reduced in the calcium deficient state, as demonstrated by the decreased slopes of the repair phase of the curves of the treated animals (Figs. 2 and 7). The study was not of sufficient length to demonstrate whether or not bone repair would have ultimately been accomplished given enough time; however, the repair curves do appear to be approaching a maximum value or steady-state plateau near the end of the 30-day period. The addition of supplemental dietary calcium in the chronic calcium-deficient state did not increase either the rate or the magnitude of the repair process. It may, in fact, retard it. Whether or not this exogenous

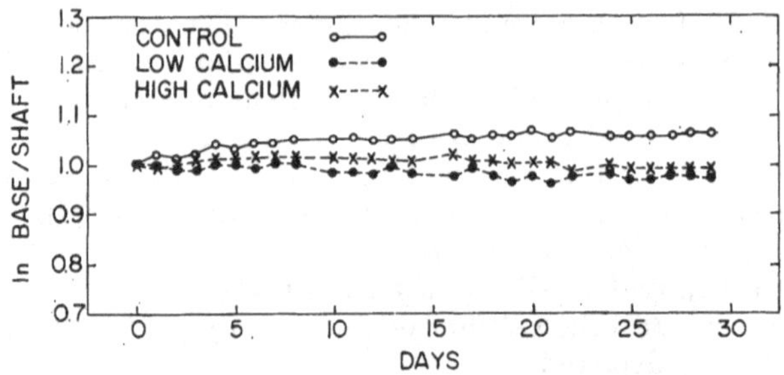

Fig. 6. Relative bone mineral mass of the control site of the control animals, low calcium animals, and animals receiving supplemental calcium (mean of standard deviations for each curve less than 0.0266).

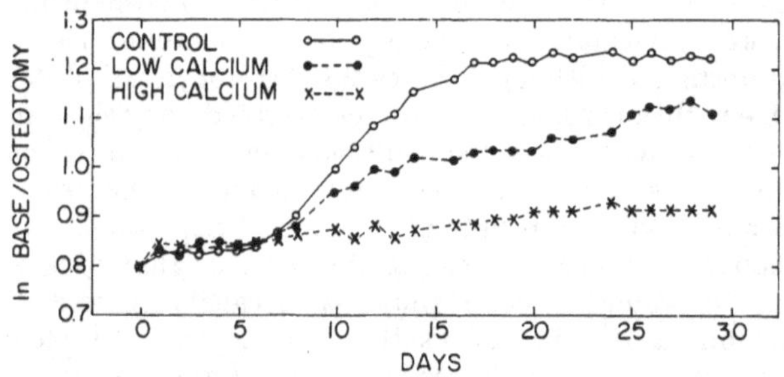

Fig. 7. Relative bone mineral mass in the healing osteotomy: control animals, animals on o w calcium rations, and animals receiving supplemental calcium (means of standard deviations of each curve less than 0.0665).

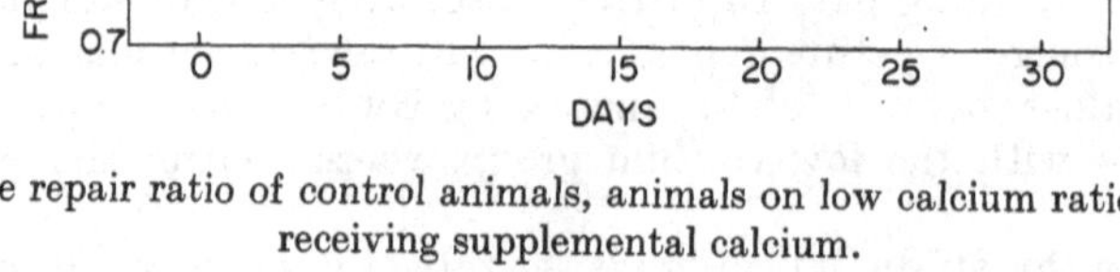

Fig. 8. Fracture repair ratio of control animals, animals on low calcium rations, and animals receiving supplemental calcium.

calcium was ultimately incorporated into the skeleton was not determined; in view of the experimental results, however, it is questionable that the non-availability of calcium was the primary cause of the lack of osseous repair. A possibility still exists that the fracture was too small to produce the stimulus necessary to initiate immediate bone repair at the expense of a calcium-deficient skeleton.

Since the animals on chronic calcium deficient diets are in negative calcium balance, the obvious lack of repair could be of clinical importance in situations where chronic negative calcium balance may exist. Space flight (weightlessness) is such a situation where a bone demineralization has been suggested [12]. The possibility exists that fractures incurred in the weightless state may be imperfectly repaired.

Acknowledgment

This work is supported by NASA Contract NAS2-5245 and NASA grant NGR-05-004-080.

References

[1] J. R. Beljan et al., Final Rep., NASA Contract No. NAS2-5245 (1971).
[2] J. R. Beljan, A. B. Hellewell and M. Goldman, Clin. Orth. **78**, 277 (1971).
[3] G. H. Bell, D. P. Cuthbertson and J. Orr, J. Physiol. **100**, 299 (1941).
[4] J. D. Benson, R. S. Emery and J. W. Thomas, J. Nutr. **97**, 53 (1968).
[5] S. H. Cohn et al., Am. J. Clin. Nutr. **21**, 1246 (1968).
[6] F. C. McLean and M. R. Urist, Bone: Fundamentals of the Physiology of Skeletal Tissue, 3rd edn, University of Chicago Press, Chicago 1968.
[7] L. C. Norris et al., J. Nutr. **102**, 1085 (1972).
[8] T. G. Taylor and J. H. Moore, Brit. J. Nutr. **10**, 250 (1956).
[9] J. M. Vaughn, The Physiology of Bone, Oxford University Press, Oxford 1970.
[10] J. R. Cameron, R. Grant and R. MacGregor, Radiol. **78**, 117 (1962).
[11] J. R. Cameron and J. Sorenson, Science **142**, 230 (1963).
[12] R. S. Hattner and D. E. McMillan, Aerospace Med. **39**, 849 (1968).

Life Science and Space Research XII — Akademie-Verlag, Berlin 1974

GRAVITATIONAL STRESS AND EXERCISE

H. Bjurstedt, G. Rosenhamer and G. Tydén

Department of Aviation Medicine, Karolinska Institutet, Stockholm, Sweden

This paper deals with the influence of G forces on certain circulatory and metabolic adjustments to dynamic exercise. Leg exercise in the sitting position increases systemic mean pressure both at normal and at increased G, thus improving G tolerance. When the force of gravity is increased to three times its normal value, leg exercise produces a much more marked increment in stroke volume than in heart rate, which contrasts with the case at normal gravity. Beta-adrenergic blockade, which strongly inhibits the heart rate response to gravitational stress, but leaves the systemic mean pressure and G tolerance unaffected, exaggerates the dominance of the stroke volume increase in exercise at increased G. With leg exercise at increasing work loads at $3G$, the O_2 uptake tends to level off and the arterial lactates to rise at a lower load than at normal gravity, indicating that the upper limit for the O_2 uptake is lowered by the G force. A progressive increase of the venous admixture in the lungs suggests that the primary limitation of the O_2 transport occurs in the pulmonary gas exchange.

I. Introduction

Although an active, healthy man spends about two thirds of the day on his feet, or at least sitting, his circulatory system is poorly designed to tolerate this continuous exposure to the earth's gravity field. The protective reflexes built into the cardiovascular system to counteract the force of gravity in the erect posture are alone insufficient. In motionless standing, therefore, hydrostatic pooling of blood into the distensible leg veins eventually brings about a serious curtailment of the cardiac output followed by a progressive fall of the systemic arterial pressure at head level and, finally, circulatory collapse [1]. However, contraction of the muscles of the calf and thigh sends blood contained in the leg veins towards the heart, and because of the numerous valves in these veins there can be no reflux during relaxation of the muscles. For the most part when erect, man is not completely still, and even small and largely involuntary contractions of the leg muscles are sufficient to prevent the cardiac output from falling to critically low levels. The leg muscles thus constitute a peripheral, accessory pump which acts as a powerful antigravity device.

Some years ago we became interested in the combined effects of exertion and increased gravitational stress. The problem had two aspects, viz. (i) how well muscular exercise could be tolerated in a high-gravity environment, and (ii) whether the normally protective effect of leg exercise on orthostatic tolerance would manifest itself also during increased gravitational stress. Several series of experiments were therefore carried out using the human centrifuge at the Karolinska Institute. The subjects were seated in the centrifuge gondola, which accommodated an electrically braked cycle ergometer with the crank axis at the level of the gondola seat. In this way the cardiovascular and respiratory effects of graded leg exercise could be investigated both at normal gravity and at $3G$ acting in the head-to-seat direction, the $3G$ level being selected since it permitted exposures of sufficient duration for steady-state measurements both in the resting and the exercise conditions. Heart rate, ECG, and intra-arterial pressure at heart level were recorded continuously, and the cardiac output was determined by the indicator-dilution technique using a remotely controlled system [2] for injection of the indicator while the subject was exercising in the spinning centrifuge gondola. Total pulmonary ventilation and O_2 uptake were measured by conventional methods. Arterial CO_2 and O_2 tensions were recorded continuously in blood sampled from one radial artery and flowing past a combined CO_2 and O_2 electrode assembly.

One early observation was that the subjective G tolerance is improved by dynamic leg exercise [3]. Whereas in the resting condition at $3G$ circulatory collapse became imminent in the majority of the subjects after 5 to 12 minutes as evidenced by loss of vision or rapid increase in heart rate, no such symptoms appeared during 13-minute exposures to the same G level when combined with continuous leg exercise at light or moderate work loads. The arterial pressure showed a rise of the same order of magnitude that occurs at normal gravity [4] and this rise can be attributed to the combined effects of increased cardiac output and vasoconstriction in non-exercising parts and is evidently the main factor responsible for the increased circulatory tolerance to G stress.

2. Cardiac Adjustment to Exercise

The mode in which leg exercise augments the cardiac output at increased G clearly differs from that at normal gravity. Thus, at 3 G cardiac adjustment to even light exercise is brought about by a much more marked increment in stroke volume than in heart rate [5]. Fig. 1 shows the changes in heart rate, stroke volume and cardiac output brought about by transition from rest to exercise at a load corresponding to 50% of the maximal O_2 uptake. It may be noted that at $3G$ in the resting condition, stroke volume is reduced by about 50%, and that with moderate exercise the stroke volume increases by no less than 81%. At normal gravity, however, the demands on the circulation in exercise are met predominantly by an increase in heart rate. This discrepancy can best be explained by the assumption that at $3G$ some of the blood pooled in the legs is redistributed towards the heart by the action of the leg muscle pump. Earlier investigations on the effects of posture shifts at normal gravity show that with light to moderate exercise in the supine position, increases in cardiac output depend almost exclusively on increases in heart rate, whereas in the upright position there is an increased contribution on the part of the stroke volume [6, 7]. Comparisons with these observations demonstrate that the response of the stroke volume to exercise becomes more marked as the magnitude of the G vector in the head-to-foot direction is increased. This also emphasizes the role of the leg muscle pump as a booster to the cardiac pump in situations where large hydrostatic pressures and pooling of blood in the lower portion of the body would otherwise tend to impair cardiac filling.

By administration of propranolol, a beta-adrenergic blocking agent, sympathetic chronotropic and inotropic stimulation of the heart can be inhibited. Propranolol strongly inhibits the heart rate response to increased G at rest and thereby leads to a more marked G-induced curtailment of the cardiac output. These effects are, however, compensated by an increase in total vascular resistance, so that the arterial pressure response and subjective G tolerance are largely unaffected by the

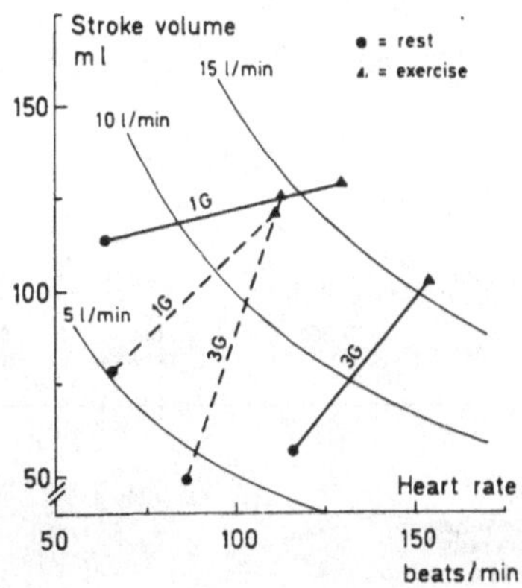

Fig. 1. Changes in heart rate and stroke volume with transition from rest to exercise at $1G$ and $3G$. Solid and broken lines refer to events before and after beta-adrenergic blockade, respectively. Curved lines indicate different levels of cardiac output. (By courtesy of *Acta Physiol. Scand.*)

drug. Beta-adrenergic blockade also reduces the heart rate response to exercise both at normal gravity and at $3G$. Fig. 1 shows that the exercise-induced increases in cardiac output at both G levels are brought about by more marked increases in stroke volume than before blockade. The difference may be explained by the assumption that the reduced heart rate response improves cardiac filling through a lengthening of the diastolic filling time and thereby augments the stroke output through the operation of the Frank-Starling mechanism.

3. Oxygen Transport

Metabolic and respiratory functions during exercise at the $3G$ level were studied at work loads up to 900 kpm min^{-1} [5, 8]. That the delivery of oxygen to working muscles is fairly well preserved at $3G$ is shown by the fact that the mean O_2 uptake during the stable state of exercise increases almost linearly with the work load up to 900 kpm min^{-1} (Fig. 2, upper graph). The values for O_2 uptake are higher at $3G$. This difference, which at any given work load amounts to 200 to 250 ml min^{-1}, can be attributed to the higher energy cost of the work of postural and respiratory muscles at high G, caused by the increased effective weight of body structures and by G-induced increase of the respiratory activity.

Inspection of individual data (Fig. 2, lower graph) reveals that with an increase of the work load from 600 to 900 kpm min^{-1} at $3G$ subjects 7 and 8 exhibited a levelling off of the O_2 uptake. Subject 3, who was able to complete the work at 900 kpm min^{-1} at normal gravity, had to abort this work at $3G$ because of exhaustion. All of these three subjects displayed considerably higher arterial lactate

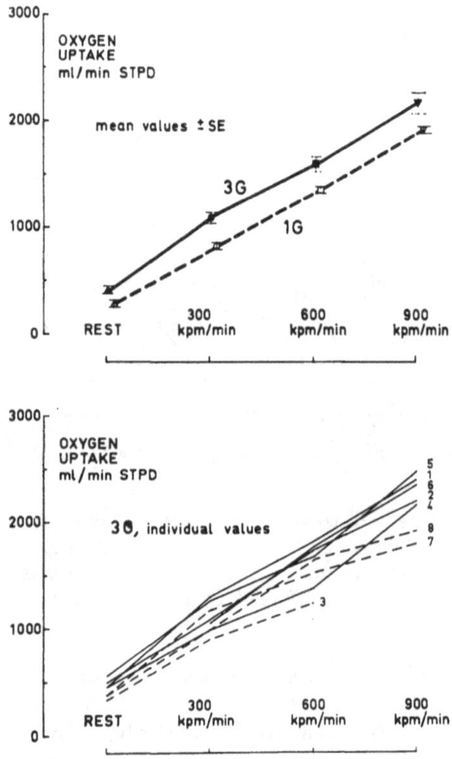

Fig. 2. Oxygen uptake (ml min^{-1}) vs external work load. Upper graph: Group means at $1G$ and $3G$ ($n = 8$, except for $3G$ at 900 kpm min^{-1}, where $n = 7$). Vertical bars indicate $\pm$SE. Lower graph: Individual data for the $3G$ level. Subject no. 3 was not able to complete the work at 900 kpm/min because of exhaustion. (By courtesy of *J. Appl. Physiol.* [8].)

values with 900 kpm min^{-1} at $3G$ (range 6.15—8.33 mEq l^{-1}) than did the rest of the subjects (range 2.0—3.68 mEq l^{-1}). The levelling off of the O_2 uptake with the linear increase of the work load, and the concomitant increase of the arterial lactate level, may be regarded as criteria that the over-all functional capacity of the O_2 transport system was being approached, and it can be inferred that this capacity was surpassed in that subject who had to abort the experiment at the highest work load.

The question naturally arises as to the nature of the disturbances along the O_2 transport line which tend to diminish the aerobic work capacity of the individual during increased gravitational stress. There is evidence that the G factor interferes both with the circulatory transport and with the O_2 uptake in the lungs.

Both during rest and exercise at $3G$ the cardiac output is considerably smaller for a given O_2 uptake than it is at normal gravity. Increased gravitational stress thus leads to a hypokinetic circulation with a widening of the overall arterio-venous O_2 difference, i.e. an increased utilization of the venous oxygen reserve.

In healthy individuals at normal gravity the pulmonary ventilation is usually not a limiting factor for the O_2 uptake. This is evidenced by the fact that, at normal gravity, the arterial O_2 tension remains unchanged or is negligibly lowered even in maximal physical exertion. In our material a change from rest to exercise at 600 kpm min^{-1} produced an insignificant fall of the arterial O_2 tension by 2 mmHg (2 torr). However, at $3G$ the corresponding fall was 13 mmHg. At the same time the alveolar oxygen tension remained elevated over corresponding values at normal gravity, resulting in a progressive widening of the alveolar-arterial O_2 difference as the work load was increased. This increase in the alveolar-arterial O_2 difference is a clear sign of impaired efficiency of pulmonary function and is predominantly due to blood flowing past lung alveoli with low ventilation/perfusion ratios. A shunt-like effect is established, since the result is the same as if a portion of the blood flowing through the lungs had done so without being arterialized. This so-called venous admixture, expressed as a percentage of the cardiac output, is small during rest at normal gravity and usually decreases with light and moderate exercise [9, 10]. However, with exercise at $3G$ the venous admixture instead increases with the work load [5] and of all the differences between the adjustments to exercise at normal and increased G this is perhaps the most striking. As to the cause of this difference, one can only speculate that the combination of increased pulmonary blood volume caused by the action of the leg muscle pump, on one hand, and the increased G load, on the other, may induce collapse of alveoli in dependent regions of the lungs where the intrapleural and intravascular pressures are high, and thus lead to an increased shunt effect.

A simple answer to the question what limits physical performance at increased G is so far not available. The present experiments have dealt with physiological adjustments to leg exercise with the G force acting in the head-to-foot direction. In this special case, our data confirm the suggestion [8] that the primary limitation imposed on the O_2 transport system occurs in the lungs.

References

[1] O. H. Gauer and H. L. Thron, in: Handbook of Physiology, Circulation Vol. III, Am. Physiol. Soc. Washington 1965 (p. 2409).

[2] G. Rosenhamer, Scand. J. Clin. Lab. Invest. **21**, 247 (1968).

[3] G. Rosenhamer, Acta Physiol. Scand. **72**, 72 (1968).

[4] D. Linnarsson and G. Rosenhamer, Acta Physiol. Scand. **74**, 50 (1968).

[5] G. Rosenhamer, Acta Physiol. Scand. Vol. **68**, Suppl. 276 (1967).

[6] S. Bevegård, A. Holmgren and B. Jonsson, Acta Physiol. Scand. **49**, 279 (1960).

[7] Y. Wang, R. J. Marshall and J. T. Shepherd, J. Clin. Invest. **39**, 1051 (1960).

[8] H. Bjurstedt, G. Rosenhamer and O. Wigertz, J. Appl. Physiol. **25**, 713 (1968).

[9] H. Bartelo, R. Beer, H.-P. Koepchen, J. Wenner and I. Witt, Pflüg. Arch. Ges. Physiol. **261**, 133 (1955).

[10] C. M. Hesser and G. Matell, Acta Physiol. Scand. **63**, 247 (1965).

Life Sciences and Space Research XII — Akademie-Verlag, Berlin 1974

MODIFICATIONS DE LA PERFUSION ET DE LA VENTILATION PULMONAIRES AU COURS DE L'IMPESANTEUR SIMULÉE

P. CALEN, R. GRANDPIERRE et A. LASNIER

University de Bordeaux, France

Les observations récemment faites au cours des vols spatiaux ont de nouveau attiré l'attention sur le rôle de la pesanteur sur la circulation sanguine et particulièrement sur la circulation à basse pression. Nous avons voulu apporter une contribution à cette étude par l'observation scintigraphique de la perfusion et de la ventilation alvéolaire chez des sujets placés au repos et à l'effort, comparativement dans des conditions normales puis en immersion. L'immersion est certes une simulation imparfaite de l'impesanteur, mais comme l'indiquait Margaria, ce qui importe au physiologiste c'est la réduction de la force que doivent exercer les structures du corps et non celle qu'exerce leur support.

Nous avons réalisé plusieurs séries d'expériences en utilisant d'abord l'iode 131 et le technetium 99 marquant les molécules de sérum-albumine pour controler les modalités de perfusion pulmonaire. Par la suite, nous nous sommes servi de xénon 133 pour étudier à la fois la perfusion et la ventilation alvéolaire. Dans la première série d'expérience, 12 sujets volontaires ont été examinés comparativement, au repos, en pesanteur normale et en immersion, étant placés en différentes positions, pendant des périodes de temps d'un minute à six heures. Dans la deuxième série, six sujets ont été observés recevant cinq injections consécutives de xénon 133, une première fois au repos, une seconde après un effort de moyenne intensité, puis après 10 minutes et 60 minutes d'immersion, et enfin, après effort en immersion depuis 60/70 minutes.

En ce qui concerne la perfusion pulmonaire, nous avons pu objectiver une hypoperfusion périphérique globale en immersion plus marquée aux sommets pulmonaires et aux bases. Cette modification est analogue à celle que l'on observe lors du passage de la position debout à la position couchée en pesanteur normale. Mais alors que cette différence de perfusion se trouve dans ce dernier cas instantanément provoquée, elle *s'établit lentement en immersion* (en 15 à 20 minutes) et reste la même après 1 heure 30. Nos observations à ce sujet n'ont pas dépassé cette durée. Ce phénomène est du même ordre que celui décrit dans les mêmes conditions par Wagner et collègues. Il est à remarquer que, pendant l'immersion, le flux sanguin cardiaque augmente jusqu'à un maximum atteint dans nos expériences vers la soixantième minute. Ce fait maintient pendant l'immersion une augmentation de la mass sanguine intrathoracique. Aussi au cours de l'effort après 1 heure 15 d'immersion, le débit cardiaque est très supérieur à celui que l'on observe au cours du même effort avant immersion.

En ce qui concerne la ventilation alvéolaire, elle se trouve au cours de nos expèriences, sensiblement augmentée dans presque tous les cas. Cet effet est net après 15 à 20 minutes d'immersion; il est plus évident encore après 60 minutes. Mais le fait qui nous semble le plus important est l'apparition en immersion de compartiments ventilatoires différents: les uns rapides (analogue à la ventilation normale), les autres plus lents. Ces derniers ne constituent du reste que 10% de la ventilation globale. Ils correspondraient à des alvéoles dont la ventilation a été réduite. Dans l'ensemble, la ventilation traduit une nette augmentation du rendement respiratoire au repos après 15 minutes d'immersion. Au cours de l'effort en immersion, la ventilation semble nettement plus efficace que lors des mêmes épreuves pratiques avant l'immersion. Il semble ainsi que les alvéoles hyperperfusées en immersion permettent des échanges gazeux beaucoup plus rapides.

L'importance des modifications de perfusion pulmonaire et de ventilation nous est apparu assez variable selon les individus. Il est à remarquer que chez les sujets «sportifs» les variations observées en immersion ont toujours été plus légères et rapidement compensées.

Remerciements

Nous remercions le Professeur P. Blanquet dans le Service duquel les études scintigraphiques ont été exécutées, surveillées et interprêtées.

Life Science and Space Research XII — Akademie-Verlag, Berlin 1974

A MONKEY METABOLISM POD FOR SPACE FLIGHT WEIGHTLESSNESS STUDIES

N. Pace, D. F. Rahlmann, A. M. Kodama, R. C. Mains and B. W. Grunbaum

White Mountain Research Station, University of California, Berkeley, Calif., USA

An experiment system is described which will permit quantitative physiological studies in 8—14 kg adult monkeys during future space flight opportunities such as the NASA Shuttle. The system comprises a fiberglass pod containing a comfortable restraint couch for the animal. The pod is divided into upper and lower halves. When the monkey occupies the couch, a rubber belly-band forms a gas seal between the upper and lower portions of the animal. The upper-pod ventilating air stream is monitored for P_{O_2}, P_{CO_2}, P_{N_2} and P_{H_2O} to permit continuous metabolic gas-exchange measurements for computation of metabolic energy expenditure. The lower pod is lined with ashless filter paper for excreta collection and entrapment of liquid urine, and is ventilated separately. At the end of experimental periods of 1 to 7 days, all excreta plus the filter paper are removed for chemical analysis. The lower-pod ventilating gas ports also allow application of lower-body negative pressure to test cardiovascular competency when desired. The upper pod is provided with automatic food pellet and drinking water dispensers for quantitative measurement of nutrient intake. Balance assessment of the following elements can be made: N, Ca, P, Mg, K, Na, Cl, S and Fe. Animals have been kept in continuous restraint in the system for as long as 30 days without apparent ill effects.

1. Introduction

It is now well established that in man chronic exposure to weightlessness results in major alterations of mineral metabolism and cardiovascular function [1, 2]. Because of the monkey's amenability to comfortable, close restraint for many days at a time [3, 4], with the attendant advantages of rigorously quantitative alimentation and excreta collection, it represents an excellent animal model for man in order to study the physiological changes which occur in space flight. It has been the experience of this laboratory that the optimal species for such studies is the pigtailed monkey, *Macaca nemestrina*, a native of Southeast Asia. Adult males of the species weigh 8—14 kg, and can be maintained in good condition indefinitely on a precisely formulated food pellet regimen which is available commercially (Purina Monkey Chow Tablets 5040, Ralston Purina Company, Richmond, Indiana 47374).

2. Monkey Pod System

With the special requirements of the proposed NASA Space Shuttle operations in mind, a lightweight fiberglass pod system which can accommodate pigtailed monkeys weighing up to 14 kg in comfortable restraint has been developed and tseted under laboratory conditions. The pod, shown disassembled in Fig. 1 and partially assembled in Fig. 2, is approximately 40 cm wide × 50 cm deep × 90 cm high when fully assembled. Including a 10 kg monkey, the assembly weighs 40 kg. As seen in Fig. 1, the pod contains a restraint couch in which the monkey is held by a fitted nylon net jacket and a pair of loose, stainless steel leg shackles, one over the thighs and one over the lower legs. A foot bar is provided which the monkey can grasp at will with his feet. The seat and lower back of the couch are padded

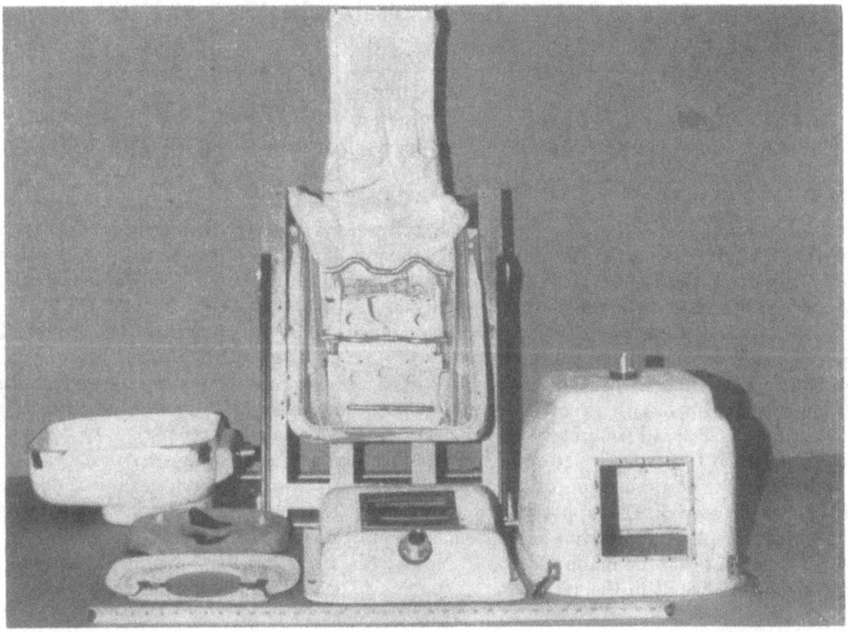

Fig. 1. Components of fiberglass monkey pod. Starting at rear center is the lower portion with restraint couch and jacket in place. Proceeding clockwise is the upper hood, the front cover of the lower portion, an arm tray, the rubber belly-band, and a center section joining the upper and lower portions of the pod.

with a viscoelastic matrix covered with silicone rubber. The couch containing a monkey can either be oriented vertically with the animal in a sitting position, or tilted backward 90° so that the animal is on its back in the astronaut launch position. Tests so far have shown that the monkeys can remain continuously in seated restraint in this couch for periods up to 30 days without apparent ill effects. It has also been possible to keep animals continuously in the supine position for 14 days in experiments analogous to bed-rest studies of humans. In these experiments clear evidence has been obtained of the development of cardiovascular decompensation, as judged from the response of heart rate and systemic arterial blood pressure either to tilting to the vertical, or to lower body negative pressure.

With a monkey in the couch, a rubber belly-band is used to form a gas seal between the upper and lower halves of the pod, which permits independent ventilation of the two halves. The lower pod is lined with sheets of ashless filter paper to soak up excreta produced by the animal. A slow stream of warm, dry air is passed continuously through the lower pod to keep the skin of the animal dry. It also serves to evaporate water from the excreta. No evidence of skin irritation has been observed, even after 30 days of continuous occupany by a monkey. The arrangement serves well to make quantitative collection of all excreta produced by an animal during an experimental period. At the end of such a period, the body surface of the lower half of the animal is carefully rinsed

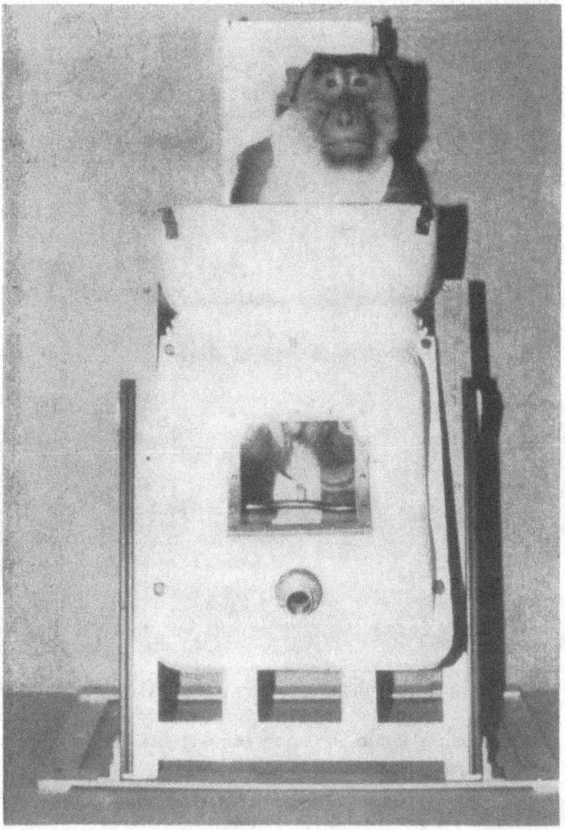

Fig. 2. Pod assembly with 10 kg pigtailed monkey in place and upper hood removed.

with distilled water into the pod. The contents of the pod, including the filter-paper lining, are transferred quantitatively to a drying tray and placed in a lyophilizer. The dried material is weighed, and extracted with petroleum ether to determine fat content. The dried, defatted excreta plus filter paper are finely comminuted in a Wiley mill and thoroughly mixed. Aliquots are then digested or ashed appropriately for measurement of the following elements: nitrogen, calcium, phosphorus, potassium, sodium, magnesium, chlorine, sulfur and iron. Four-day collection periods have been found to be convenient, although the duration can readily be varied from 1 to 7 days. Consecutive collection periods are also entirely feasible.

The lower pod with the rubber belly-band in place can be used for the application of lower body negative pressure when desired. Pressure differentials up to 80 torr have been used successfully with these monkeys in the pod.

With the upper half of the pod in place, a constant inflow of fresh air at rate of 8—12 liters/min is sufficient to maintain a P_{IO_2}* of approximately 150 torr and a P_{ICO_2} of 5 torr, as well as to provide for removal of water evaporating from the upper part of the animal. Mass flow of the exhaust gas from the upper pod is measured continuously, as is P_{O_2}, P_{CO_2}, P_{N_2} and P_{H_2O}. In the present laboratory

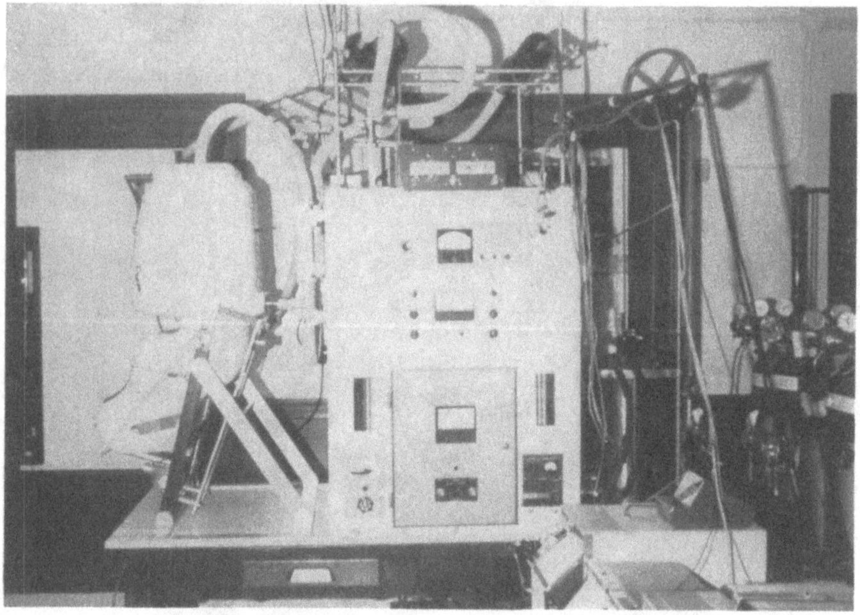

Fig. 3. Complete pod assembly attached to respiratory gas analysis instrumentation. The latter is enclosed in a constant-temperature box kept at 35 °C to prevent water vapor condensation in the hood exhaust-gas line.

version of the pod system these measurements are accomplished by standard instrumentation, as shown in Fig. 3. It is expected that the NASA respiratory-gas mass spectrometer developed for the Skylab Program would be used for the measurements in space-flight experiments with the monkey pod. The upper pod also incorporates a simple food-tablet dispenser, not shown in the figures, which is operated by the monkey *ad libitum* through manipulation of a spring-loaded handle. If desirable, more elaborate performance-task feeders can be used. An *ad libitum* automatic water-dispenser nozzle, which is activated by 10 torr of oral suction, is also provided within the upper pod. These devices permit continuous recording of feeding and drinking activity, as well as accurate measurement of food and water intake during an experiment. The food-tablet and drinking-water reservoirs are outside the pod, and are readily accessible for replenishment.

---

* P_{IO_2} = partial pressure of oxygen in inspired air.

3. Typical 4-Day Pod Experiment

The uppermost graph in Fig. 4 summarizes the mean 24-hour values for O_2 consumption rates and CO_2 production rates measured during a 4.1-day pod experiment with a 12 kg male pigtailed monkey. The O_2 consumption rate $\dot{V}_{O_2}$ ranged from 100.6 to 115.4 liters/24 hr, with an overall mean value of 108.5 liters/24 hr, during the 4.1-day experiment. Similarly, CO_2 production rate $\dot{V}_{CO_2}$ ranged from 91.2 to 102.0 liters/24 hr, with an overall mean value of 98.5 liters/24 hr.

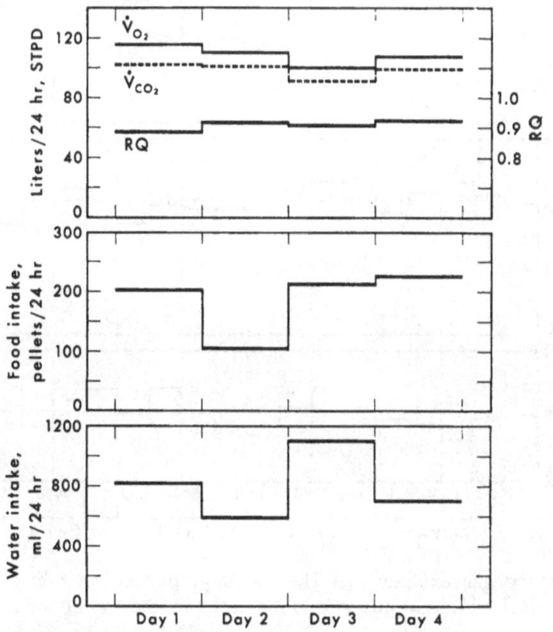

Fig. 4. Respiratory gas exchange, food intake and water intake of a 12 kg pigtailed monkey during a 4.1-day pod experiment, plotted as 24-hr values.

It may be noted that the respiratory quotient *RQ* displayed 24-hour mean values ranging from 0.88 to 0.92, with an overall mean value of 0.91 for the entire 4.1 days. The middle graph in Fig. 4 indicates the number of food tablets ingested per 24 hours during the experiment. The animal showed a transitory reduction in food intake on the second day in the pod, but then seemed to stabilize at a higher level for the remaining days. The animal consumed a total of 789 tablets during the 4.1 day period, or a mean intake of 192 tablets/24 hr. The tablets weigh 0.79 g each, so that the animal ingested an average of 152 g of food per day. The bottom graph in Fig. 4 shows the 24-hour water intake during the experiment. It may be seen that the animal drank from 600 to 1100 ml of water per 24-hr period. He consumed a total of 3210 ml in 4.1 days for a mean value of 780 ml/24 hr.

All of these data can be resolved into shorter time periods, as desired. For example, Fig. 5 shows the respiratory gas exchange data plotted in 6-hr periods during the same 4.1-day experiment. There is a suggestion of a 24-hr periodicity in the O_2 consumption and CO_2 production rate data, but the total number of days of measurement is too small to permit frequency analysis. On the other hand, it is evident that the animal was in a generally stable state during the entire 4.1-day period of the experiment. The food and water intake patterns are

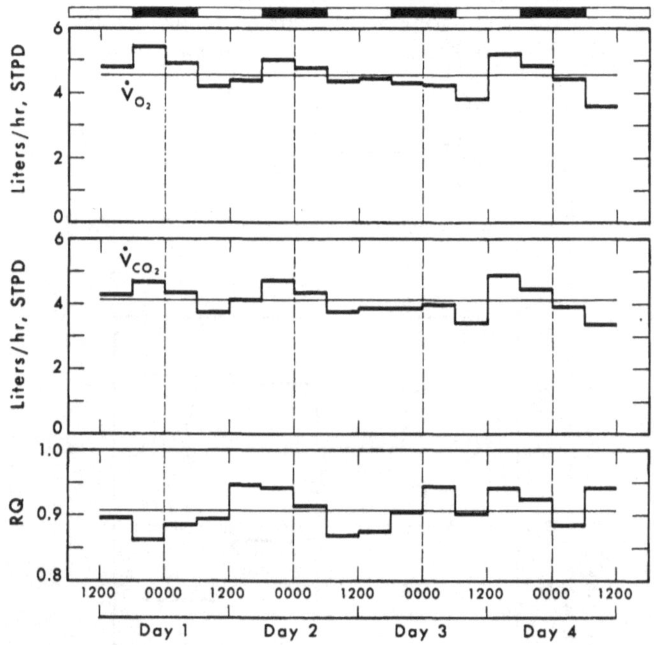

Fig. 5. Respiratory gas exchange of the monkey, plotted as 6-hr values. The pod light: dark illumination cycle is indicated at the top of the figure.

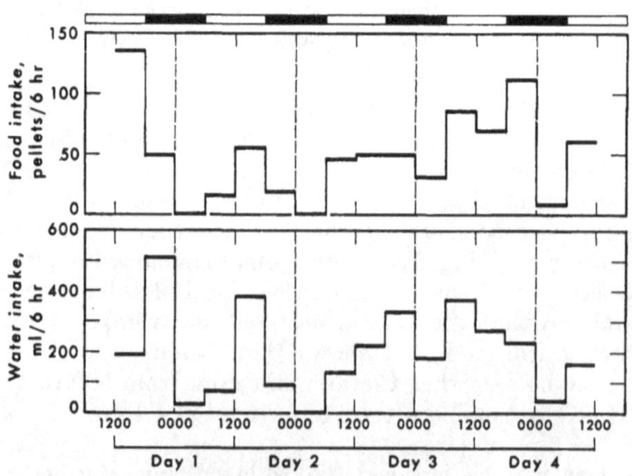

Fig. 6. Food and water intake patterns of the monkey, based on 6-hr intervals.

similarly plotted at 6-hr intervals in Fig. 6. A somewhat unexpected finding was the continued feeding and drinking activity of the animal during the 12-hour dark period from 1800 to 0600, and particularly in the hours before 2400. It is tempting to speculate that the tendency of man to conduct midnight raids on the refrigerator may represent a basic primate behavioral trait, and should therefore not evoke too many guilt feelings.

Table 1 gives the composition of the food tablets provided to the monkey during the 4.1-day experiment. The values are based both on data provided by

Table 1

Total Caloric Value of the 625 g of Food Tablets ingested by the Monkey Subject of a 4.1-day Pod Experiment

Tablet constituent	Composition (%)	Total intake (g)	Caloric value (kcal)
Carbohydrate	60.8	380.0	1520
Protein	22.0	137.5	550
Fat	2.0	12.5	113
Water	8.0	50.0	0
Fiber	3.0	18.8	0
Minerals	4.2	26.2	0
Total	100.0	625.0	2183

the Ralston Purina Company and on independent analyses made in our laboratory. It may be seen that the diet formulation is one relatively high in carbohydrate and protein content, and relatively low in fat content, and altogether provided 2183 kcal of metabolizable energy to the animal during the 4.1-day period. A theoretical *RQ* may also be computed for the metabolic utilization of the 625 g of tablets eaten during the experiment, and a value of 0.92 is obtained. This agrees very well with the mean *RQ* value of 0.91 calculated from the total O_2 consumption and CO_2 production measurements for the 4.1-day period as indicated earlier. The total caloric value of the food ingested by the animal during the experiment, 2183 kcal, may be compared with the metabolic energy expenditure during the same period as computed from O_2 consumption rate, using the standard conversion factor of 4.85 kcal/liter O_2 consumed. During the 4.1-day period the animal consumed 445 liters of O_2, which corresponds to an expenditure of 2158 kcal. Thus, the animal was in nearly perfect energy balance. This conclusion is supported by the observation that the body weight of the animal was 12.35 kg when he was put into the pod, and 12.34 kg when he was removed 4 days later. As further verification, according to Kleiber [5], the metabolic rate for mammals may be computed from the relationship $M = 72\,W^{0.75}$ where M is the resting metabolic rate in kcal/24 hr and W is the body weight in kg. In the present instance, M is computed to be 490 kcal/24 hr, or 2009 kcal in 4.1 days. This agrees within 7% with the energy expenditure of 2158 kcal computed from the O_2 consumption measurements.

It was also possible to carry out an analysis matrix on the total excreta collection from the experiment, for the 9 elements mentioned earlier. However, the procedure is a new one and must be regarded as preliminary in its present form. The results of the elemental analysis on the total excreta, together with the total intake of the nine elements during the same period, are shown in Table 2 primarily for illustrative purposes. For six of the elements, the animal seemed to be essentially in zero balance for the 4.1-day period. However, the results indicate an apparent positive balance for nitrogen, phosphorus and iron.

In the case of nitrogen there is growing evidence [6, 7] that a significant fraction of the body turnover may involve loss through conversion of amino nitrogen to

Table 2

Metabolic Balances for 9 Elements in the Monkey Subject of a 4.1-day Pod Experiment, estimated from a Total Food Tablet Consumption of 625 g and Analysis of the Total Excreta Collection

Element	Total intake (g)	Total excreted (g)	Net retention (+) or loss (−) (g)
Nitrogen	22.21	18.18	+4.03
Calcium	5.69	5.46	+0.23
Potassium	4.44	4.58	−0.14
Phosphorus	3.19	2.33	+0.86
Chlorine	2.94	2.82	+0.12
Sodium	1.41	1.56	−0.15
Magnesium	0.90	0.78	+0.12
Sulfur	0.365	0.348	+0.017
Iron	0.206	0.127	+0.079

N_2, especially on excessively high protein diets such as the present one. Thus, estimation of nitrogen balance without taking this loss into account would indicate an apparent positive nitrogen balance. Another possible source of error might be the conversion by bacterial decomposition of significant quantities of excreta amino nitrogen to ammonia in the course of several days. However, we have quantitatively collected the ammonia appearing in the lower-pod ventilating air, and after 4 days less than one per cent of the nitrogen turnover could be accounted for in this form. It appears, then, that a diet formulation containing substantially less protein would be of considerable advantage in assessing nitrogen balance in such studies, while at the same time supplying the minimum daily requirement of essential amino acids.

The most likely explanation for the apparent positive phosphorus and iron balance in this particular experiment is analytical artifact. As already stated, the methodology for making rigorously accurate simultaneous determinations of all nine of these elements in heterogeneous biological samples is still undergoing development in our laboratory, but on the basis of the results to date there is every expectation that the requisite level of accuracy and reproducibility is within close reach. The primary problem has been one of achieving representative sampling of the large bulk of material involved, rather than one of the individual analytical methods themselves.

4. Summary

In summary, the monkey pod system described permits the continuous, quantitative definition of metabolic activity in a good animal model for man under carefully controlled conditions for a number of days at a time. It also possesses structural and operational characteristics which are well suited to its application in studies of the physiological effects of weightlessness during spacecraft flight.

Acknowledgments

This work is supported under NASA Grant NGL 05-003-024, as a part of the research program of the Ames Research Center, Moffett Field, California.

References

[1] L. Lutwak et al., J. Clin. Endocr. **29**, 1140 (1969).
[2] C. A. Berry, in: Bioastronautics Data Book, 2nd edn, National Aeronautics and Space Administration Publication SP-3006, 1973 (p. 349).
[3] N. Pace et al., Aerospace Med. **35**, 118 (1964).
[4] D. F. Rahlmann et al., Lab. Animal Care **14**, 125 (1964).
[5] M. Kleiber, Fire of Life, Wiley, New York 1961.
[6] G. Costa et al., Nature **218**, 546 (1968).
[7] J. H. Cissik et al., J. Appl. Physiol. **32**, 155 (1972).

Life Sciences and Space Research XII — Akademie-Verlag, Berlin 1974

THE ROLE OF GRAVITY IN THE PHYLOGENY OF STRUCTURE AND FUNCTION IN ANIMAL SENSORS OF SPATIAL ORIENTATION, AND THEIR PREDICTED ACTION IN WEIGHTLESSNESS

YA. A. VINNIKOV*

Sechenov Institute of Evolutionary Physiology and Biochemistry, USSR Academy of Sciences, Leningrad, USSR

The evolution of the structural, functional and cytochemical organization of the gravity receptor which determines a body position in the gravitational field of the earth by means of muscular regulation was traced both invertebrates and vertebrates, using electron microscopic and histochemical methods. In the course of evolution of vertebrates, the specialized gravity receptor-statocyst which, as a rule, consists of primary sensory cells and supplies otoliths, is formed. In vertebrates, there exists a vestibular apparatus made up of secondary sensory cells and also having otoliths. The receptor cells, both of statocysts and the vestibular apparatus, are supplied with special antennas (kinocilia and stereocilia). Deviation of the antennas stimulated by displacement of the otoliths resulting from locomotor activity of animals leads to excitation of the receptor cells. When exposed to a modified gravitational field (linear accelerations of $10g$, for 3 min), the receptor cells of the vestibular apparatus, in all classes of vertebrates, show progressive changes in RNA content and protein synthesis (increase followed by decrease) which return to normal only after 12 days. Thus, immediate transfer of animals and man from acceleration to weightlessness appears to be a reason for movement disease.

The above consideration showed the need for an experiment in which an animal (with its vestibular apparatus) which had not undergone previous accelerations, would be exposed to weightlessness. Frog embryos, *Rana temporaria*, at the stage preceding the organogenesis, when the vestibular apparatus and other organs were lacking, were chosen as a suitable subject. Frog embryos at the stage of an early gastrula were placed in a special container Emkon aboard the Soyuz 10 spacecraft. After short accelerations, they were exposed to weightlessness for 44 hours. The embryos were allowed to continue to develop to the stage of early tail bud. The experimental embryos showed normally developed acoustic vesicles and vestibular ganglia. Clear differentiation of the receptor cells with antennas (kinocilia and stereocilia) was found in the acoustic vesicles. Thus, in weightlessness, vestibular apparatus develops just as well as in the gravitational field of the earth. However, only a much longer stay in weightlessness conditions will indicate whether there are any changes in the structural, cytochemical and functional organization of vestibular apparatus.

The similarity in the structural, functional and cytochemical organization of the gravity receptor in vertebrates and invertebrates appears to allow the prediction of the behaviour of the gravity receptor as a whole, and of its receptor elements, both in normal and changed gravitational fields. The first attempts were carried out only on the vestibular apparatus of vertebrates.

---

* The co-authors of the report are O. G. Gazenko, L. K. Titova, A. A. Bronstein, T. P. Tsirulis, R. A. Pevzner, V. I. Govardovskii, F. G. Gribakin, L. P. Pal'mbakh, M. Z. Aronova, A. L. Mashinskii, V. P. Ivanov, T. A. Kharkeevich and G. A. Pyatkina.

1. Introduction

For orientation in the gravitational field of the earth most animals are equipped with a "body position sensor", that is, a special otolith organ; this has reached varying degrees of perfection during evolution, but it is built on the same basic principle in all organisms. The important point is that the equilibrium organ consists of two parts: (i) a "test mass", i.e. an otolith (which sometimes has a non-biological origin) that can move more or less freely within the organ and (ii) a system of receptors capable of perceiving the position or movement of this mass within the organ. Any deviation in the body position is accompanied by a displacement of the "test mass" (otolith, otoconia, otolith membrane) which stimulates a corresponding group of receptors. The signal generated by these receptors is processed by the central nervous system which directs command signal to muscles in order to correct the body position. In this sense we can say that the organism possesses special "gravity receptors". When using this term, it is important to remember that the cells of the equilibrium organ do not by themselves perceive gravitation; they can perceive only the magnitude and direction of that force which is exerted on them by the "test mass" under the effect of the earth's attraction. The aim of the present report is to follow the evolution of the gravity receptor, as revealed by electron microscopy and histochemistry, in principal types of invertebrates and vertebrates and to establish the principal structural and cytochemical changes resulting from stimulation of this organ in "modified" gravitational fields (the effect of linear and angular accelerations as well as that of weightlessness).

2. The Principles of Structural, Cytochemical and Functional Organization of the Gravity Receptor of Unicellular, Invertebrate and Protochordata Organisms

When dealing with species of various types of invertebrates it is possible to trace how natural selection is able finally to create, by the "trial and error" method, very perfect gravity receptors, similar in different invertebrates in its structural, cytochemical and functional organization, and capable of perception of body position with reference to gravitational field.

In this respect the data on the statocyst-like organelle of protozoa [1, 2] are of particular importance. The displacement of concretions due to movement of a protozoon is believed to be detected by the cytoplasmic membrane of the wall of the organelle and the signal is transmitted through the cytoplasm to motile cilia that in turn regulate the body position. If we can show that the statocyst-like organelle of protozoa performs the role of a receptor, then it is evident that in the course of evolution this function has been structurally designed in an appropriate manner in unicellular animals.

The quest for a better design of gravity receptor is best seen in Coelenterata and Ctenophora. Indeed, coelenterates display all variations in gravity receptors — from a small receptive field covered with lithocytes (i.e. cells containing a concretion in the cytoplasm), to a lithostyle or rhopalia, (which is some kind of elastic pendulum with its principal mass formed by lithocytes). If the lithostyle is immersed in a pit (which sometimes appears closed) this is often considered to be a statocyst, although a real statocyst has a much more complicated organization. Any change in the body position in the gravitational field results in deflection of the lithostyle loaded with lithocytes. This deflection is detected either by surrounding receptor cells or by receptor cells which lie on the lithostyle itself.

The organization of the aboral organ of Ctenophora shows considerable progress (Fig. 1). Here, for the first time in evolution, we can find receptor cells each

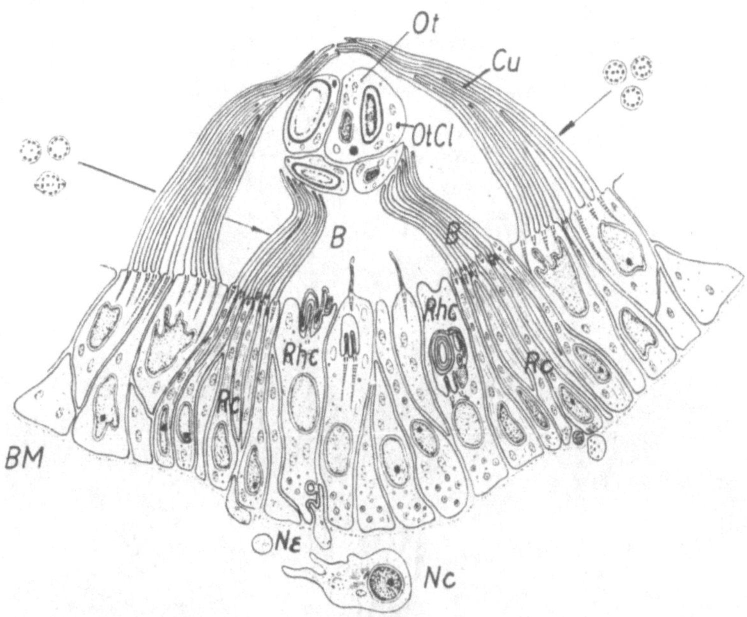

Fig. 1. Diagram of the ultrastructural organization of the gravity receptor (aboral organ) in Ctenophora. RC, receptor cells of the gravity receptor; PhC, photoreceptor cells; B, balancers; Cu, cupule; Ot, otolith; NC, nerve cells; NE, nerve endings; BM, basal membrane.

supplied with one cilium (containing 9 pairs of peripheral fibrils and two central ones) and a free otolith produced by special cells—otoconiablasts. Displacement of the otolith caused by the deflection of the body axis from the vertical line causes deflection of the bundle of cilia that are combined to form "springs" or "balancers" (Fig. 1). Excitation of the receptor cells due to the deflection is transmitted, it is believed, through synapses to the nervous system. Ctenophora are known to have no other sense organs, so that integration of practically all the functions of these animals is performed by the aboral organ, through the nervous system.

Other forms of sense organ are found in the first representatives of Deuterostomia, namely in flatworms. Turbellaria possess a special organ, the statocyst, which is a vesicle filled with liquid. According to our present information it consists of only one cell containing a cavity within which the oscillating lithocyte is bound by a slender pedicle. Statocysts of turbellaria lack free otoconia, whereas those are already found in the aboral organ of Ctenophora. The lithocyte, first originated in coelenterata, plays the part of a "test mass" in these animals. In nemerteans the organization of the statocysts is apparently similar to that of turbellaria. Annelids have well developed statocysts which may communicate with the environment by means of a narrow duct. Otoconia in annelids may be brought in from the environment or they may be formed by the secretory activity of the

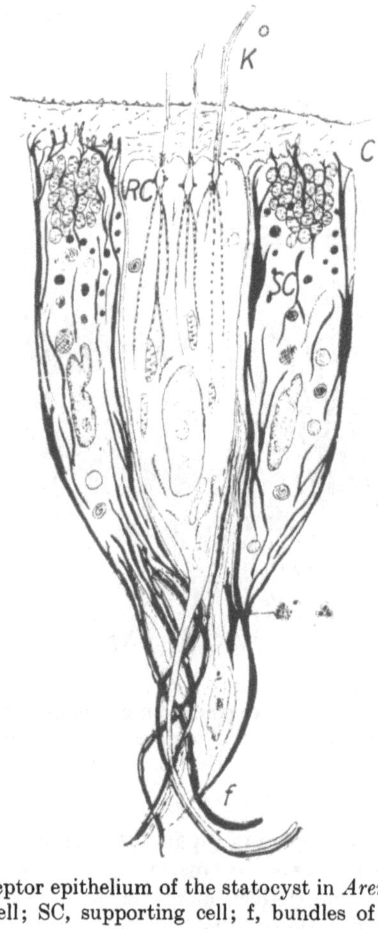

Fig. 2. Diagram of the receptor epithelium of the statocyst in *Arenicola marina*. K, kinocilia; C, cuticle; RC, receptor cell; SC, supporting cell; f, bundles of fibrils in supporting cells.

cells. The electron microscopic data obtained in our laboratory show that it is the statocyst of annelids where we notice for the first time the appearance of primary sensory cells equipped with cilia containing $9 \times 2 + 2$ fibrils (Fig. 2). The deflection of the cilia caused by movement of the otolith results in excitation of the receptor cell and transmission of the signal to the nervous system of the worm.

The most perfect form of organization of a statocyst which is transformed into a true equilibrium organ is observed in arthropods, and in particular in higher crustaceans. Statocysts of higher crustaceans consist of an invagination of the integument that openly communicates with the environment by means of a cleft. Only in lower crustaceans are the statocysts completely separated from the integument. However, in both higher and lower crustaceans the chitinous cuticular lining of the statocyst changes during every moulting period, and then the vesicle-shaped statocysts of lower crustaceans are temporarily also in contact with the environment. During moulting, the statocyst lining is removed together with the sand particles (otoliths) enclosed in it. After moulting new sand particles are introduced either by the claws or by repeated immersion of the head into sand. It was on statocysts of the crayfish that Kreidel's classical experiment was performed [3], in which iron filings instead of sand were offered to the animal after moulting to fill up its statocyst, and the animal was then exposed to the action of a magnet; the magnetic field then simulated the effect of the gravitational field. Receptor cells of the statocyst of crustaceans are typical primary sensory cells. The bulk of the cell is located on the base of the statocyst which is parallel to the earth's surface, i.e. lies horizontally. The axon of the cell extends to the central nervous system. According to our data as well as those of Schöne and Steinbrecht [4] the peripheral process is a modified cilium containing 9 pairs of peripheral fibrils. This process penetrates into a chitinous sheath which has the shape of a hair. The number of hairs may be as great as 400–500. The tips of chitinous hairs are attached to the otolith membrane. Change in body position with respect to the gravitational field is perceived by the peripheral processes of the receptor cells which are thus the gravity receptors. The other hairs situated on the lower surface of the statocyst are not attached to the otolith membrane. They are stimulated by streams of liquid and perceive angular acceleration. Some hairs have been proved to respond to viibration. Thus, a very effective type of gravity

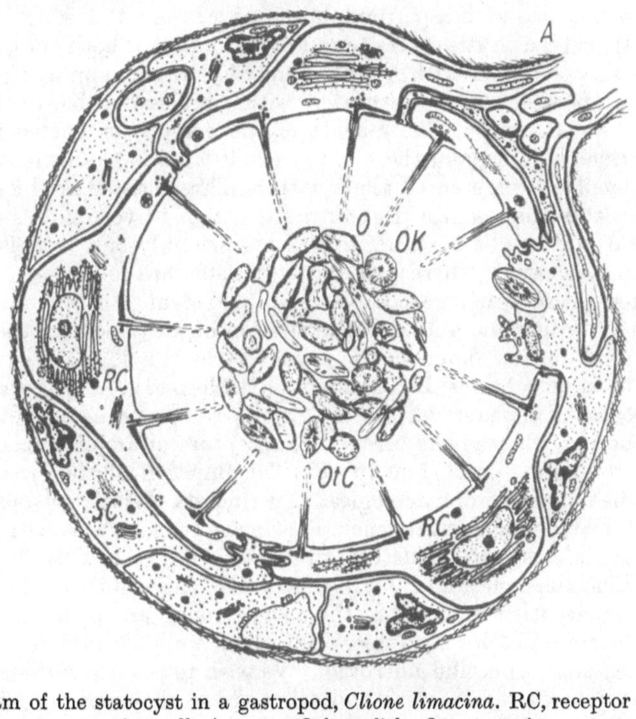

Fig. 3. Diagram of the statocyst in a gastropod, *Clione limacina.* RC, receptor cell; SC, supporting cell; A, axon; Otl, otolith; Ot, otoconia.

receptor has been evolved in higher crustaceans; this model represents a universal sense organ of both static and dynamic equilibrium. A novel device is shown by the chitin lining of the statocyst: in this particular case, contact between the otolith membrane and the peripheral processes of the receptor cells required the introduction of an additional transformer in the form of the chitinous hair-cap.

The statocysts in molluscs are far more perfect (Fig. 3). They are found in species belonging to all classes of mollusc. In gastropods, the statocyst is a small round sac filled with endolymph and containing, apparently, a free otolith. A limited region of sensory cells, a macula, has been found in the statocyst of Pterotrachea. The sensory cells of the macula are primary sensory cells and are equipped with hairs. As our electron microscopic study has shown, the hair-tipped giant primary sensitive cells of the statocyst of the fresh water gastropod mollusc, *Planorbis corneus*, are typical kinocilia. A similar structure is exhibited by the cells of bivalve molluscs, in the statocysts of which, however, maculae are absent. The mechanism of contact between kinocilia and otolith (otoconia) in the statocysts of gastropod and bivalve molluscs is still not clear. Are the otoliths attached to the distal ends of kinocilia or are they free to move in the endolymph? But one way or another the deflection of the kinocilia of receptor cells is achieved by displacement of otoliths when the position of the mollusc body changes in the gravitational field. There is reason to believe that efferent innervation, which regulates the functions of the receptor cells, originated early in the statocyst of gastropods.

Cephalopod molluscs have reached the highest level of organization. A study of their gravity receptors is of particular interest in view of the jet propulsion they use to move. In the octopus statocysts are provided not only with an endolymphatic cavity but also with a perilymphatic space; the latter is not found in the squid (Fig. 4). The basic structures found in statocysts of both octopus and squid are the maculae equipped with otoliths, and the cristae, appearing for the first time in invertebrates, which are coated with cupules capable of deflection by streams of the endolymph. The macula of the octopus is located in a vertical plane; the principal macula of the squid is also located in the same plane but two additional maculae form a small angle with it. Three cristae are located in three mutually perpendicular planes. In contradistinction to vertebrates two of them are horizontal and one is vertical. In the octopus only one anticrista protrudes into the statocyst cavity whereas eleven anticristae are found in the squid, and their arrangement represents a complex spatial system. It is assumed [5, 6] that the presence of anticristae allows the regulation of movement of the endolymph when the animal moves about, in the same way as this is achieved in the semicircular canals of vertebrates. In a series of physiological studies it was established that the maculae of cephalopods are the gravity receptors, i.e. receptors ensuring static equilibrium, whereas the cristae are receptors of angular acceleration, i.e. they ensure dynamic equilibrium [6–8]. The function of the cristae has been recently studied by electrophysiological experiments (Wolff, personal communication). Both maculae and cristae include primary sensory cells with a very complex cytochemical and ultrastructural organization (Fig. 5A, B). These cells are tipped with kinocilia containing $9 \times 2 + 2$ fibrils. The distal terminations of the kinocilia are apparently connected to the otolith membrane or to the cupula. The kinocilia are surrounded by 2–5 microvilli which we identified, by analogy with vertebrates, as stereocilia-like microvilli. We wish to point out the fact that the receptor cells in both the maculae and cristae have their apical surfaces sloped. The basal pedicles of kinocilia are regularly oriented in the direction of the slope.

Such an orientation may indicate morphological and functional polarization of the cells. It may be of interest that invertebrates other than molluscs lack polarization of this kind, although little is known in this respect about arthropods. Vertebrates, as we shall see below, have their receptor cells also regularly polarized at their tips but the polarization is achieved in a different way. Unipolar neurons, also having complicated ultrastructural and cytochemical organization, are located under the receptor cells of the cephalopod statocyst. Such neurons (which are not found in statocysts of other molluscs) have an integrative function providing communication between groups of the receptors. However no experimental data are available on this. The bodies and axons of these neurons may reach giant dimensions and may connect with a layer of nerve fibres adjacent to the central processes of the receptor cells. As electron microscopic observations and

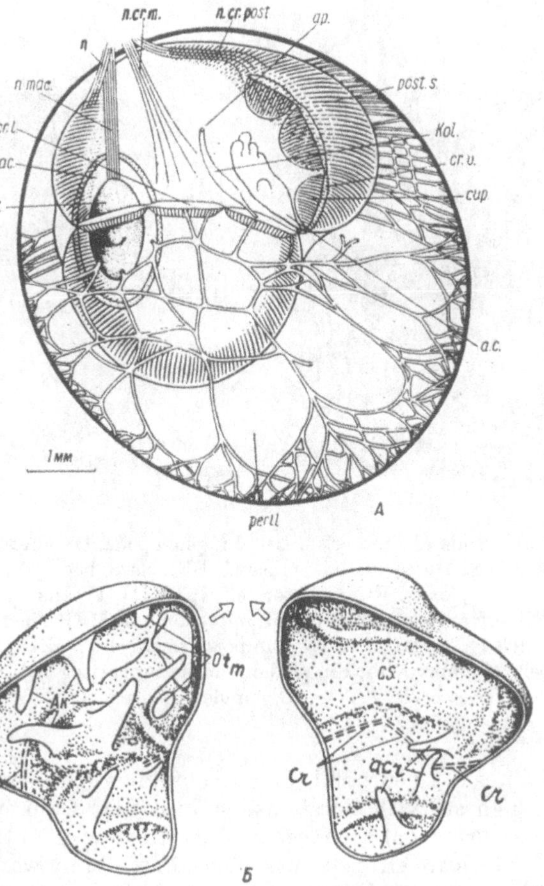

Fig. 4. Diagram of the statocyst in Octopoda (*Octopus sp.*) (A) and Decapoda (*Omnatostrephes sloanei pacificus*) (B) [7]. A: n, nerve, n.cr.m., nerve of medial crista; n.cr.post., nerve of posterior crista; ap., apex of the Kölliker canal; post.s., sinus posterior; Köl., Kölliker canal; cr.v., ventral crista; cup., cupule; ac., anticrista; peril., perilympha; cr.l., lateral crista; ot., otolith; mac., macula; n.mac., macular nerve. B: acr., anticrista; otm., otolith of macula; CS., statocyst cavity; Cr., crista.

cytochemical data have shown, present among these fibres are fibres filled with synaptic vesicles and exhibiting high acetylcholinesterase activity. Between these fibres synaptic contacts are often observed which indicates the possibility of functional connections between axons of the neurons and central processes of the receptor cells, as well as cholinergic fibres. These facts make us consider that the

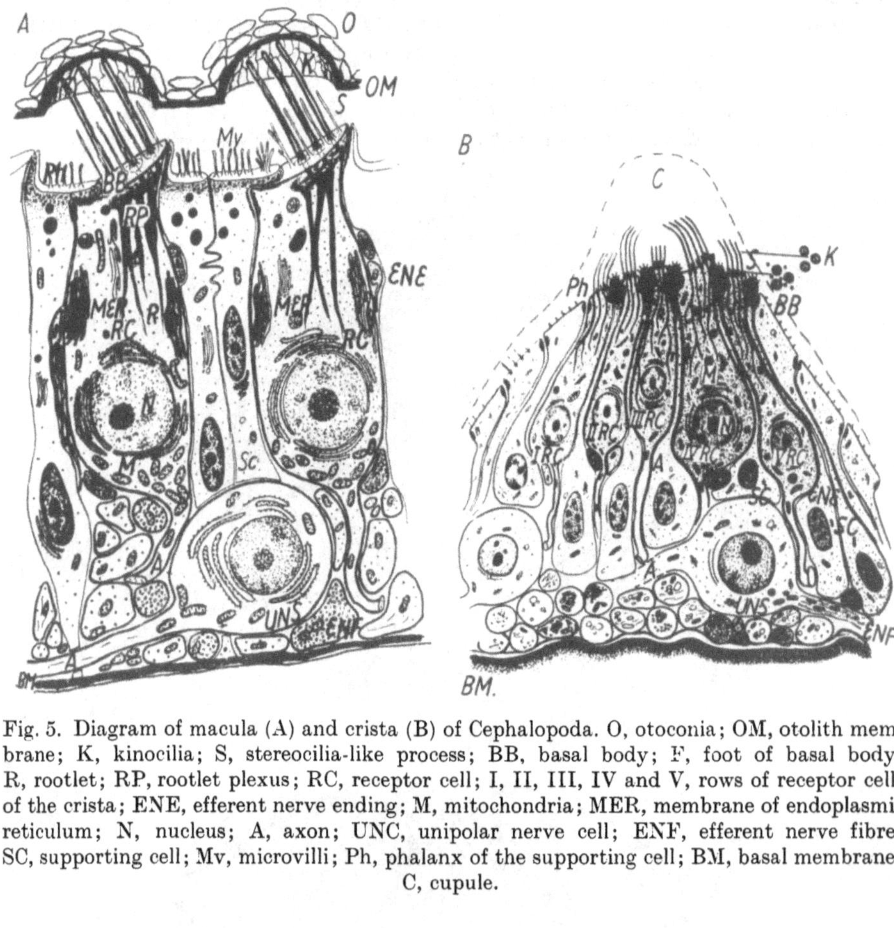

Fig. 5. Diagram of macula (A) and crista (B) of Cephalopoda. O, otoconia; OM, otolith membrane; K, kinocilia; S, stereocilia-like process; BB, basal body; F, foot of basal body; R, rootlet; RP, rootlet plexus; RC, receptor cell; I, II, III, IV and V, rows of receptor cells of the crista; ENE, efferent nerve ending; M, mitochondria; MER, membrane of endoplasmic reticulum; N, nucleus; A, axon; UNC, unipolar nerve cell; ENF, efferent nerve fibre; SC, supporting cell; Mv, microvilli; Ph, phalanx of the supporting cell; BM, basal membrane; C, cupule.

layer of nerve fibres is a neurophil. Fibres containing numerous synaptic vesicles and exhibiting high acetylcholinesterase activity may form synaptic contact at the base and side surfaces of the receptor cells. In view of all these facts, we conclude that these fibres are efferent ones. Thus, in the samy way as in vertebrates, the function of the receptor cells in statocysts of cephalopods is apparently regulated by means of a feedback On the basis of all the facts presented above, the statocysts of cephalopod molluscs can be considered as quite effective devices, namely as equilibrium organs, which in many respects (including their connections with the central nervous system and other sense organs) are just as effective as the equilibrium organs of vertebrates.

We shall not give here special consideration to the statocysts of echinoderms and urochordata, which, unfortunately, have not been much studied. Although these animals are deuterostomians, they are provided with primitive multicellular or unicellular gravity receptors of the lithocyte typo found in coelenterates and turbellaria. On this basis we can not assume that the gravity receptor of urochordata is the initial form from which the gravity receptor of vertebrates has evolved. Thus, an analysis of the evolution of the gravity receptor of unicellular organisms, invertebrates and urochordata shows how, by the "trial and error" method, natural selection has finally produced the most rational structure for the gravity receptor in all independently arising phyletic lines; practically the same elements are utilized, namely a primary sensory cell equipped with kinocilia, and a "test mass" in the form of an otolith or otoconia. From the axamples of evolution of the gravity receptor of invertebrates we can see that with the development of the motor activity in animals this receptor is transformed from an organ that arose as an organ of strictly static equilibrium into an organ capable of regulating not only the behaivour of the animal in the gravitational field but also coordination of its movement in three-dimensional space. This second function reaches a specially advanced stage of development in vertebrates.

3. Principles of the Structural, Cytochemical and Functional Organization of the Gravity Receptor in Vertebrates

The gravity receptor in all vertebrates (cyclostomata, fishes, amphibians, reptiles, birds and mammals) consists of the utricular and saccular maculae and the semicircular canals. The organization of the receptor structures in the labyrinth, including the utricle of lower vertebrates (fishes and amphibians) is based on the so-called type II cells; in addition to these cells, type I cells are found in higher vertebrates, starting with reptiles (Fig. 6). Individual button-shaped nerve endings (boutons) approach the type II cells; type I cells are immersed in special, cup-shaped nerve endings. Both types of cells are crowned with a bundle of hairs consisting of some tens of stereocilia and a single kinocilium having a polar orientation with respect to the stereocilia. Type I cells appear for the first time in higher vertebrates in connection with more active movements of the animals in the gravitational field; this required a more universal and reliable transmission of information from the receptor cells to the central nervous system. At the same time, evolutionary transformations at the cellular level did not affect at all the organization of hairs and their connections with the otolith membrane, which in type I cells does not differ in any way from the phylogenetically earlier type II cells. Therefore, primary physical transformations, associated with reception of the earth's gravitational field, seems to have been resolved by natural selection at the dawn of evolution of multicellular animals. Later, when new types of gravity receptor appeared in new animal types, including vertebrates, the same structure which adequately senses the gravitational field was inevitably evolved again. Numerous investigators have repeatedly emphasized the fact that successful morphological, physiological or biochemical solutions of certain biological problems once they were evolved by natural selection in unicellular animals were then preserved in evolution in multicellular organisms, where they were refined and adapted to more elaborate functions [9].

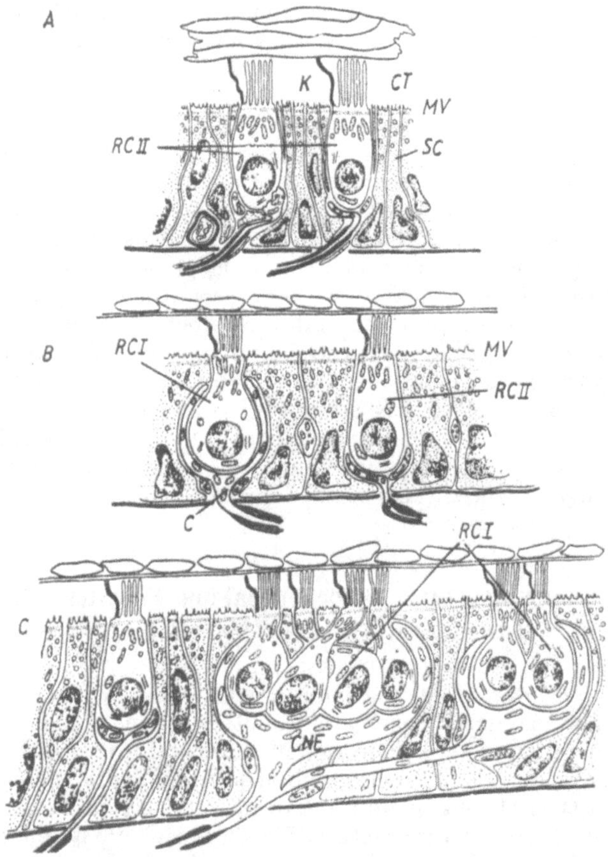

Fig. 6. Diagram of the evolution of structural organization of utricle in vertebrates. A, fishes; B, mammals; C, bird. RC I and RC II, type I receptor cell and type II receptor cell; SC, supporting cell; K, kinocilium; CT, stereocilia; CNE, chalyx-like nerve ending; MV, microvilli

In all vertebrates a change in the body position in the gravitational field, that causes shearing movements of the utricle otolith membrane, results in bending of a bundle of the receptor cell hairs. Bending of the stereocilia toward the kinocilium is followed by a depolarization of the cellular membranes; if the stimulus is aimed toward the opposite side hyperpolarization sets in [10]. In the first case the activity of impulses in afferent nerve fibres increases, in the second it decreases [11, 12]. If the bundle of the stereocilia does not connect with the inner surface of the otolith membrane (and as bristles of the brush can only stick in its clefts during movement of the membrane) the distal part of the kinocilium, according to our work, penetrates deeply between the cross-beams of the otolith membrane and seems to attach to it (Fig. 7). A contraction of the kinocilium (due to molecular interactions between ATP and ATP-ase of the kinocilium fibrils) seems to control the extent of reversible displacement of the otolith membrane.

The mechanochemical processes taking place in stereocilia are more difficult to understand; according to Dohlman's calculations [13, 14], the total area of the

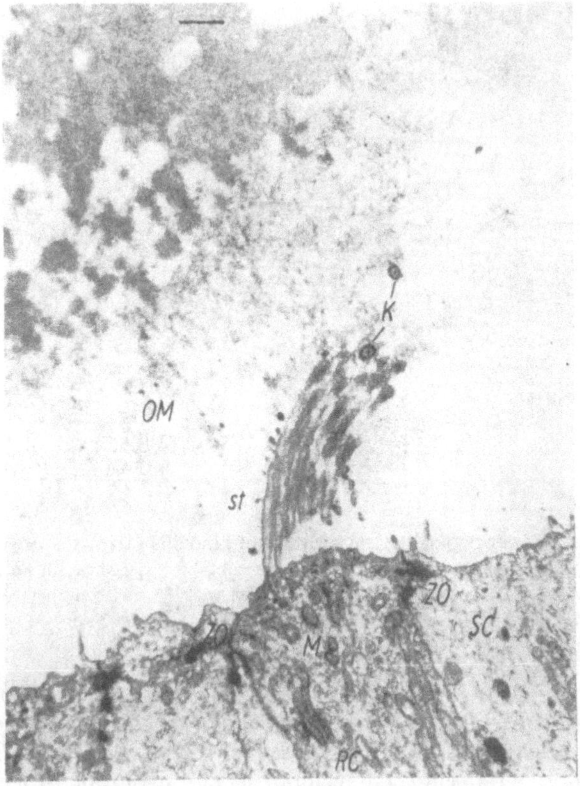

Fig. 7. The utricular macula of a turtle, *Testudo horsfieldi*, after expose to acceleration of 10*g* for 3 minutes. Bending of the stereocilia towards the kinocilium is shown. ×8000. RC, receptor cell; M, mitochondria; SC, supporting cell; K, kinocilium; St, stereocilia; OM, otlolith membrane; ZO, zonula occludens.

membranes of the stereocilia is about 2000—4000 μm^2 while the remaining plasma membrane of the cells covers an area of only 200—300 μm^2. The role of the stereocilia in the receptor cells of maculae and cıistae at the molecular level will, no doubt, be clarified in the near future when cytochemical and electron microscopic methods are improved. At present we can say little about the significance of the stereocilia.

Unlike the cristae of the semicircular canals (the kinocilia of which have quite similar orientation, i.e. in the horizontal ampulla the orientation of kinocilia is utriculopetal, but is utriculofugal in both vertical ampullae) the receptor cells with similar polarization in the utricular macula are assembled in groups, which together form a four-zone mosaic. Owing to this mosaic, details of which differ in various representatives of vertebrates (Fig. 8) only a definite group of cells is stimulated during shearing of the otolith membrane. According to physiological data obtained by Magnus [15] and especially by Szentagothai [16] and his associates who also divided the utricle into four quadrants, it follows that a definite group of muscles (which are involved in positioning the body and are regulated

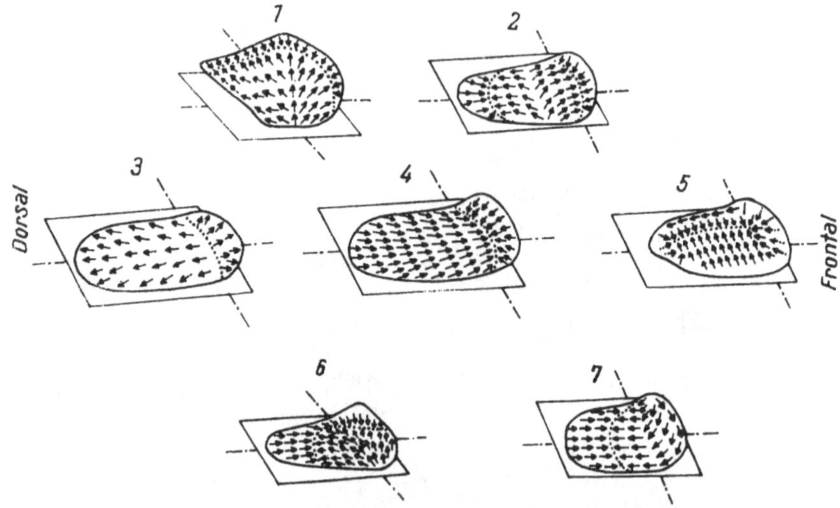

Fig. 8. Diagram of the morphological polarization of kinocilia (arrows), concerning the bundle of stereocilia, in the receptor cells of utricle in vertebrates. 1, fishes; 2, amphibia; 3, reptile (Testudo); 4, reptile (Varanus); 5, pigeon; 6, guinea pig; 7, white mouse. The most distinct lines of polarization are shown by dotted line.

by the function of the utricular macula) correspond to each portion of this mosaic or quadrant. At one time we suggested [17] that the difference in the nature of the organization of the distribution mosaic of receptor cells in the maculae of different vertebrates reflects the development and evolution of the musculature. We believe that this viewpoint is supported by comparisons of the mosaic of the utricular maculae in various animals. In birds, as we have seen, this mosaic is more complex than that in mammals; this complexity is probably associated with the peculiarities in the development of their flight musculature. On the contrary, in reptiles, and especially in the turtle, this mosaic is considerably simpler, and this should reflect a reduction of complexity of their musculature. In amphibians the mosaic, although in part similar to that of fishes, is somewhat more complex than in fishes, which, of course, reflects the appearance of limbs in these animals and their emergence from water on to land. Thus, the character of the distribution pattern of receptor cells in the maculae, depending upon the polarization of kinocilia, is a reflection of ecological transformations, induced by a change in the behaviour of animals in the earth's gravitational field.

4. Change in Labyrinth Structures under the Condition of the Changed Gravitational Field

4.1. Linear Accelerations

We were able to establish [18] that during stimulation by radial acceleration of $10g$ for 3 minutes substantial changes in the subcellular and cytochemical organization of the nucleus and cytoplasm take place in the receptor cells of the utricle, and also (during appropriate stimulation) in the receptor cells of the sacculus and semicircular canals of the labyrinth. These changes manifest themselves in a decrease in the cytoplasmic RNA content, in a migration of the nucleolus towards the nuclear membrane, and in the transfer of ribosome RNA of the nucleolus from the nucleus to the cytoplasm, where RNA induces formation of a limited sector of the endoplasmic reticulum. These processes can be demonstrated in all vertebrates studied but are particularly distinct in guinea pigs (Fig. 9).

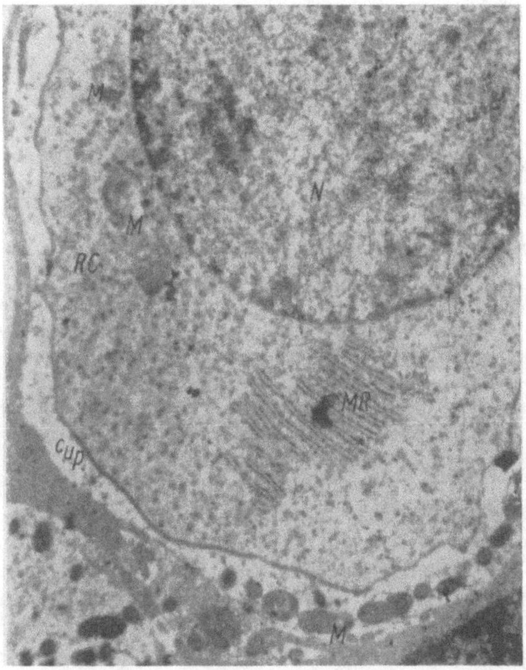

Fig. 9. The formation of apparently spiral-like rough endoplasmic reticulum in the receptor cells of guinea pig utricle exposed to acceleration of $10g$ for 3 minutes. $\times 13500$. RC, receptor cell; N, nucleus; M, mitochondria; MR, membranes of rough endoplasmic reticulum; cup., cup-shaped nerve ending.

Apparently, a directed leakage of RNA of nucleoli from the nucleus into the cytoplasm can also take place spontaneously, without exposure to an experimental stimulus, thus reflecting some sort of rhythmic entry of RNA from the nucleus into the cytoplasm [19]. Protein repair takes place with the aid of this mechanism, consequent on the considerable consumption of structural and functional proteins (enzymes) in the stimulated receptor cells during their intensive work under acceleration.

We have shown that following the restoration of cytoplasmic RNA a process of increased protein synthesis begins in the receptor cells of the utricle; after reaching a certain maximum value the protein level drops again almost to normal and then increases again, forming a multipeak curve with the highest point observed 48 hours after exposure to a single effect (Fig. 10). Repeated overloads in

1, 2, 8 hours and 12 days apparently exert an unfavourable effect, causing a disruption or even cessation of the protein synthesis.

A study of the utricle of animals in a state of rest and after exposure to overloads has shown that mitochondria of receptor cells may come into the most intimate contact with presynaptic membrane and can abut tightly against this

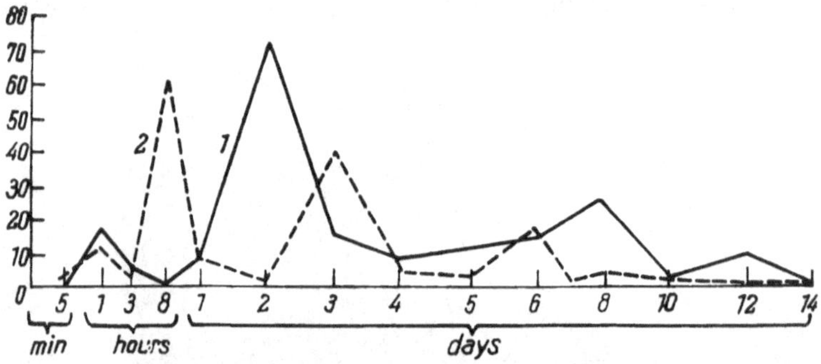

Fig. 10. Graph showing arithmetical means variation curves of the optical density of protein in the basal part of receptor cells (cytophotometric data) at different stages of recovery after a single stimulation (solid line) and repeated stimulations (broken line). The abscissa shows the time which had elapsed after the time of single rotation. Ordinate, relative units of optical density from the millimetre scale of a microphotometer (Mph-4).

membrane. As a rule, a movement of mitochondria is followed by their swelling in both cap-shaped and button-shaped nerve endings. This phenomenon can be observed in the receptor cells of the utricle rather distinctly. The phenomenon can be produced by experimental acceleration, which, in our opinion, is convincing morphological evidence for the hypothesis that mitochondria take part in transmission of a nerve impulse in the receptor elements of the gravity receptor. This conclusion is supported by the data obtained in our laboratory on cytochemistry of oxidative enzymes in the utricle of vertebrates. The activity of the oxidative enzymes in the nerve endings of the gravity receptor was shown to change during various stages of its functioning.

According to our electron microscopic and cytochemical data, acetylcholinesterase of the labyrinth is involved in the transmission of afferent inpulses from the receptor cells to the sensory nerve fibre of the neuron situated in Scarp's ganglion. Acetylcholinesterase activity, especially high in teleosts [20, 21] was demonstrated in the neurons of Scarp's ganglion. The enzymic activity of acetylcholinesterase can also be detected in efferent nerve endings of the labyrinth ([13, 14, 22—26] and our own observations). We consider as strange the statement that only efferent impulses can be transmitted with the aid of a cholinergic system; one cannot exclude the possibility that adrenergic fibres can also take part in the transmission of a nerve impulse [27].

Thus, there is a clear pattern in the structural, cytochemical and functional organization of the receptor cell of the vestibular apparatus in vertebrates, as well as in the changes which take place at both cellular and ultrastructural levels of organization as a result of stimulation. These start with the deviation of stereocilia and kinocilia under the shearing forces which arise when the otolith membrane slides in one or other direction, and which are caused by a change in the body position in the gravitational field. This mechanism is believed to constitute the basis of the function of the gravity receptor of all vertebrates.

4.2. Development of the Vestibular Apparatus (the Labyrinth) of the Frog, *Rana temporaria*, under Conditions of Weightlessness

In a previous section we pointed out that the receptor cells of the vestibular apparatus in guinea pigs exposed to acceleration of $10g$ for 3 minutes show very intensive changes. The changes which occurred in both the ultrastructural organization of the receptor cells and synthesis of the nucleic acids require at least 12 days for their repair. Hence, transfer of animals and men from normal gravity

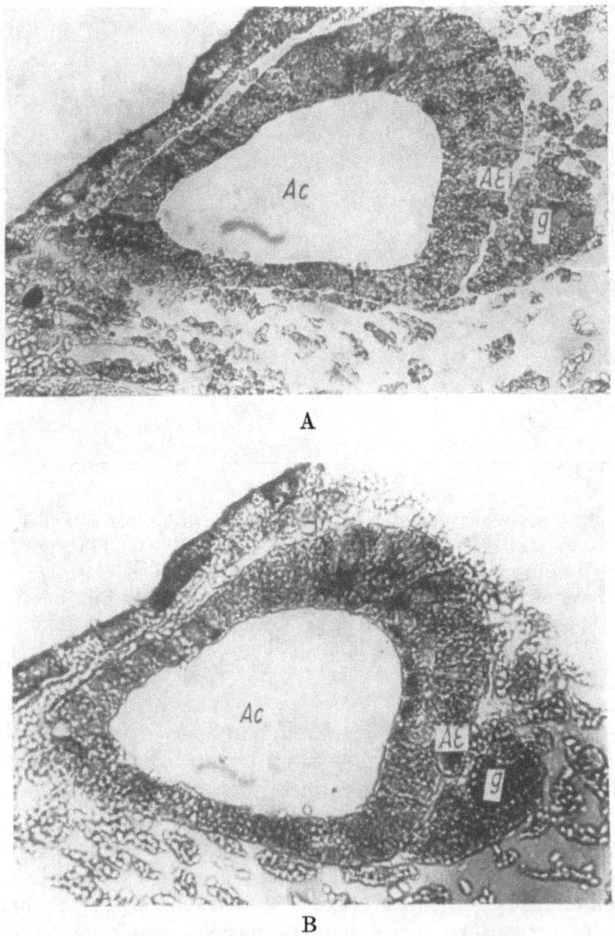

Fig. 11. Acoustic vesicle of Rana temporaria at a stage of tail bud. ×100. A — control; B — experiment. Ac — a cavity of acoustic vesicle; AE — epithelium of acoustic vesicle and macula communis; g — initial stages of differentiation of VIII ganglion.

to the conditions of weightlessness undoubtedly impose upon the vestibular apparatus new functional requirements that are as yet unclear. If the sensory organ is incapable of fulfilling these requirements this seems to lead to motion disease. We, therefore, have experimented on the vestibular apparatus of animals under conditions of weightlessness without previous acceleration. This was compared with that of embryos and animals in the normal gravitational field of the earth. Fertilized amphibian (*Rana temporaria*) eggs were used.

We present below the data obtained during experiments carried out by the crew of the spacecraft Soyuz 10. These were carried out on eggs at the late blastula

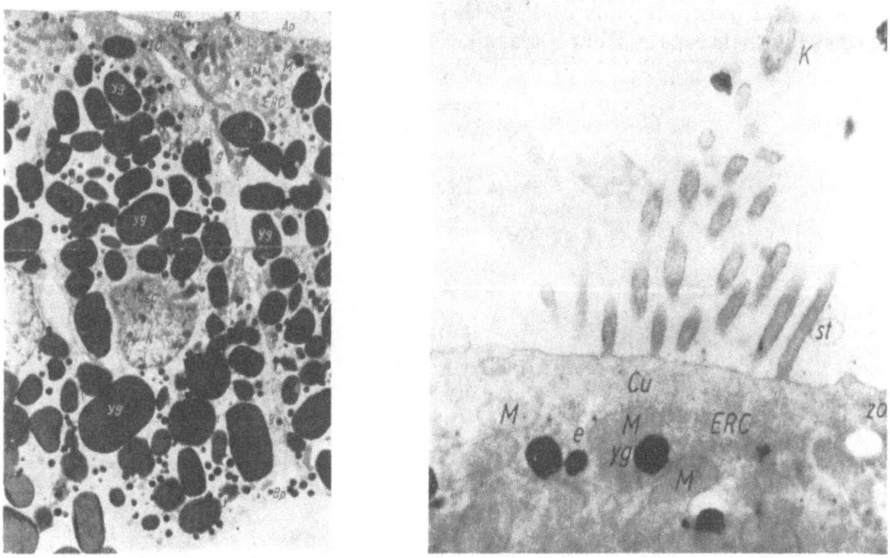

a b

Fig. 12. The wall of the acoustic vesicle of *Rana temporaria* (experiment). a, receptor epithelium, × 3800; b, apex of differentiating receptor cell, × 20000. IEC, undifferentiated embryonic cell; ERC, embryonic receptor cell with developing kinocilium; Ap, apical part of the cell; Bp, basal part of the cell; AC, cavity of acoustic vesicle; Bm, basal membrane; yg, yolk granules; P, pigment; M, mitochondria; N, nucleus; g, intercellular gap; ZO, zonula occludens; K, kinocilium; St, stereocilia; Cu, cuticle.

and early gastrula stages which were artificially fertilized on the land 43 hours before take-off. Embryos that developed under conditions of weightlessness from the gastrula stage (after brief acceleration during take off) reached the tail-bud stage just as the controls on the earth, and at this stage they were fixed. If any changes were produced by short-time acceleration and vibration they were recovered in the course of following development under conditions of weightlessness. As we know, amphibian embryos of these stages are characterized by rather wide ability of repair [28]. The structure of the acoustic vesicles which developed under conditions of weightlessness did not differ from control ones.

Under conditions of weightlessness not only does formation of the acoustic vesicle (Fig. 11 A, B) take place, but also the process of an organ differentiation begins: a typical macula communis begins to develop and a rudiment of VIIIth ganglion forms. At the same time, in some parts of the macula communis the processes of cellular differentiation begin, i.e. formation of individual receptor and supporting cells (Fig. 12a, b, 13). The apical parts of the receptor and supporting cells differentiate. Before the apical parts of the cells become free from yolk and pigment granules the mitochondria concentrate just at the place that testifies to an increase of energetic processes in the differentiated parts of the cells. It is

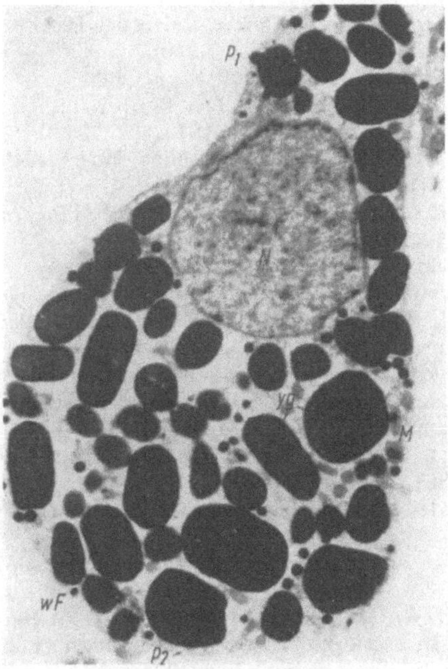

Fig. 13. Neuroblast of VIII ganglion (experiment), × 6600. P_1, P_2, opposite poles of neuroblasts; NF (written wF in diagram), beginning of nerve process formation; N, nucleus; yg, yolk granules; M, mitochondria.

evident that during the following stages of embryogenesis these cells will reach the places of the future utricular, saccular maculae, lagena and cristae of semicircular canal ampullae and will turn into definitive receptor cells. Further investigations are needed to show what will happen to the vestibular apparatus under conditions of weightlessness for a longer period.

In the controls as well as in the experiment the cellular differentiation in the embryonic VIII ganglion leads to formation of the pear-shaped bipolar neuroblasts with two processes located polarly (Fig. 17). Thus amphibian embryos under the conditions of weightlessness for two days develop as well as those under control conditions. Weightlessness practically does not affect the development of amphibian embryos at the stages of neurula and tail bud, and formation of the organs and beginning of the cellular differentiation, including that of the future vestibular apparatus.

References

[1] V. A. DOGIEL, Arch. Protistenkunde **58**, 319 (1929).
[2] V. A. DOGIEL, YU. I. POLYANSKII and E. M. KHEISIN, General Protozoology, Nauka, Moscow, Leningrad (in Russian) 1962.
[3] A. KREIDEL, Versuche an Krebsen. Sitzberich. Acad. Wiss. Wien. Abt. 3, **102**, 149 (1893).
[4] H. SCHÖNE and R. A. STEINBRECHT, Nature **220**, 184 (1968).
[5] J. Z. YOUNG, Proc. Roy. Soc. B **152**, 3 (1960).
[6] J. Z. YOUNG, A Model of the Brain, Oxford University Press, Oxford 1964.
[7] B. B. BOYCOTT, Proc. Roy. Soc. B **152**, 78 (1960).
[8] S. DIJKGRAAF, Publ. stazione. Zool. Napoli, **32**, 64 (1961).
[9] E. M. KREPS and N. A. VERZHBINSKAYA, Izv. Akad. Nauk SSSR, **6**, 855 (1959).
[10] O. LÖWENSTEIN and Y. WERSÄLL, Nature, **184**, 1807 (1959).
[11] O. LÖWENSTEIN and A. SAND, J. Physiol. (London) **99**, 89 (1940).
[12] D. TRINKER, Pflug. Arch. Physiol. **264**, 351 (1957).
[13] G. F. DOHLMANN, in: Neural Mechanisms of the Auditory and Vestibular Systems, Springfield, Ill. 1960 (p. 258).
[14] G. F. DOHLMANN, Confin. Neurol. (Basel), **20**, 169 (1960).
[15] R. MAGNUS, Körperstellung. Experimentellphysiologische Untersuchungen über die einzelnen bei der Körperstellung in Tätigkeit tretenden Reflexe, über ihr Zusammenwirken und ihre Störungen, Berlin 1924.
[16] J. SZENTAGOTHAI, Role of individual receptors of the labyrinth during orientation of the eyes and head in space, Nauka, Leningrad (in Russian) 1967.
[17] YA. A. VINNIKOV, in: Nervnaya Kletka, Leningrad 1966 (p. 7).
[18] YA. A. VINNIKOV, O. G. GAZENKO, K. K. TITOVA, I. V. OSIPOVA and A. A. BRONSTEIN, Dokl. Akad. Nauk SSSR, **153**, 450 (1963).
[19] A. A. NEIFAKH, in: Kletochnaya Differentsirovka i Induktsionnye Mekhanizmy, Nauka, Moscow 1965 (p. 38).
[20] L. K. TITOVA and YA. A. VINNIKOV, in: Evolutsiya Funktsii, Nauka, Leningrad 1964 (p. 249).
[21] YA. A. VINNIKOV, O. G. GAZENKO, L. K. TITOVA, A. A. BRONSTEIN, I. V. OSIPOVA, V. I. GOVARDOVSKII, M. Z. ARONOVA and I. L. ZHINKIN, in: Funktsionalnaya Evolutsiya Nervnoi Sistemy, Nauka, Leningrad 1965 (p. 170).
[22] G. F. DOHLMAN, J. FARKASHIDY and F. SALONNA, J. Laryngol. Otol. **72**, 984 (1958).
[23] G. ROSSI and G. CORTESINA, Ann. Laryngol. Otol. Rinol. Faringol. **61**, 296 (1962).
[24] G. ROSSI and G. CORTESINA, Panminerva Med. **4**, 478 (1962).
[25] G. ROSSI and G. CORTESINA, Y. Laryngol. Otol. **77**, 202 (1963).
[26] G. ROSSI, G. VOENA, S. BUONGIOVANNI and G. CORTESINA, Acta Oto-laryngol. **58**, 159 (1964).
[27] H. SPOENDLIN, in: The Vestibular System and its Diseases, Univ. of Pennsylvania Press (p. 39).
[28] J. S. HUXLEY and G. R. DE BEER, The Bases of Experimental Embryology Nauka, Moscow (in Russian) 1936.

Life Sciences and Space Research XII — Akademie-Verlag, Berlin 1974

HISTOLOGICAL STUDIES ON THE VESTIBULAR ORGAN OF FROG EMBRYOS AND LARVAE AFTER SIMULATED WEIGHTLESSNESS

W. Briegleb

DFVLR, Institut für Flugmedizin, Bonn, Bad Godesberg, FRG

The ontology of gravity receptors is discussed. As well as phylogenetic aspects of the question, we may ask whether the different gravity receptors in plants and animals derive from the same source; did they evolve from a general sensitivity of the living cell or do they represent parallel inventions. One may also ask whether during ontogeny of the receptor it receives information from gravity or whether the development of function of the organ is an autonomous process. In many cases ontogeny shows traces of phylogeny. Statoliths, e.g. the differentiating statolith membranes of the frog embryo, look somewhat like sedimentational structures. Furthermore the amphibian embryo in its relatively early stages takes up a special position in relation to the direction of gravity. Which mechanism is involved here? The materials studied were frog embryos and larvae which developed within a fast rotating clinostat. The results may help to answer the question as to whether the differentiation of the frog statolith membrane is an autonomous process.

1. Introduction

In earlier publications we showed that the phenotype of frogs did not change if the development from fertilization of the egg until the early larvae took place under conditions of simulated weightlessness [1]. A special question concerns the development of the statolith membranes (statoliths) in the vestibular organ. Statoliths in general look similar to sedimentational structures. Will their shape during embryogenesis be influenced by changes in acceleration? We did not find any lasting anomalies of locomotion in larvae and adults of two species of anurans whose embryos developed under conditions simulating zero gravity [2].

2. Method

There are great difficulties in making histological studies of the embryology of vertebrate statoliths. In histological surveys of the vestibular apparatus reported in the literature one may find the extracellular structures appear as coagulated masses or residues. Statolith membranes are at least shrunken [3—5]. We attempted to use the cryotome for preparation and developed the following method. Embryos or larvae of *Rana temporaria* were narcotized, placed on a cryostat-cryomicrotome and then snap-frozen. The cryotome was placed in a cold-chamber (2—4 °C).

Instead of using the tissue sections for observation the cut surfaces of the object itself were observed and photographed [6]. In this way essentially undisturbed views of the statolith apparatus were obtained. The objects were late embryos and larvae of stages 10—16 of normal development of *Rana temporaria* [7]. They came from a fast running clinostat (90 rev min^{-1}) in which they had developed from fertilization of the egg until stage 10 or 11. Others were rotated further after

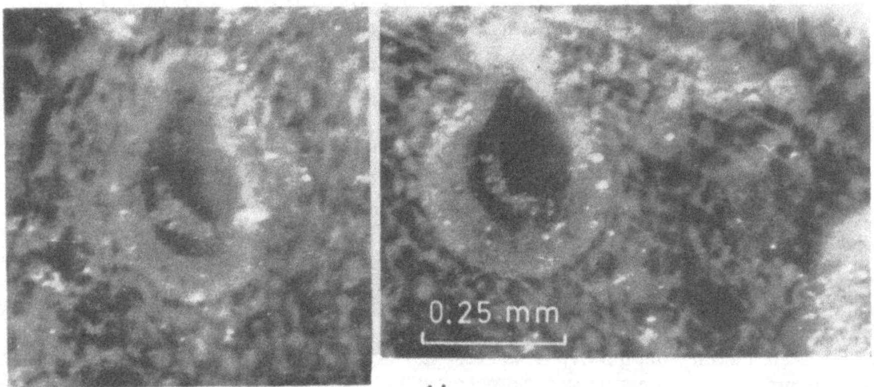

1a 1b

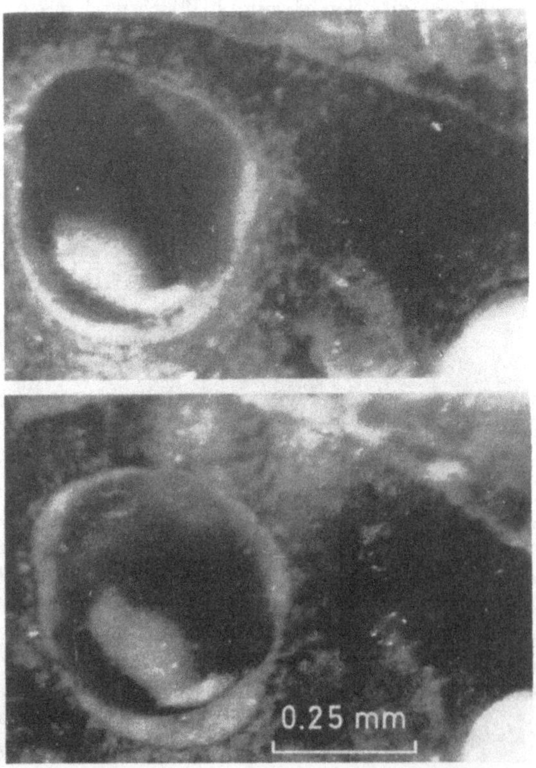

2a 2b

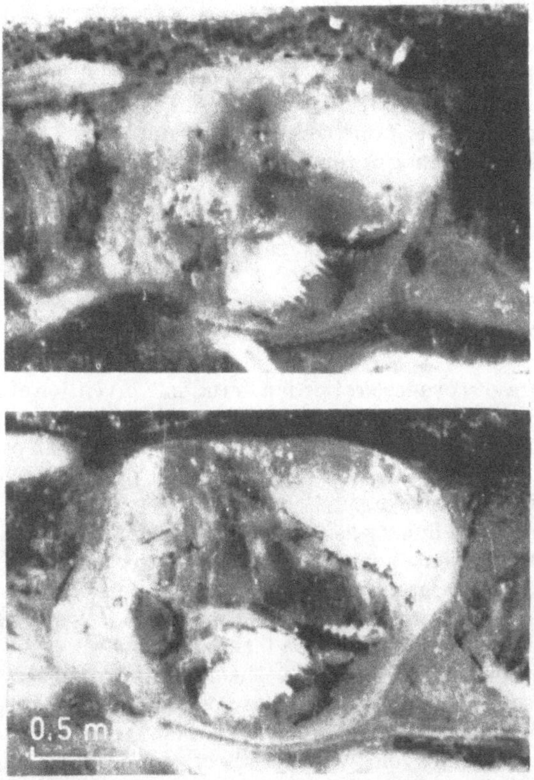

3a

3b

Figs. 1—3. Lateral views of the vestibular organ of test and control embryos and larvae of *Rana temporaria*. The test animals were rotated at 90 rev min^{-1} at 15°C. Figs. 1—3 show stages 10, 12—13 and 15—16 of the normal table of development [7]. 1a, 2a, 3a are controls; 1b, 2b, 3b are test animals. In 2b the animal was rotated in two different rotors until preparation; in 3b rotation was continued only until hatching.

changing the rotor. For details of the method see also [1, 8]. Other animals were allowed to develop for about three months under normal conditions after removing them from the rotor.

3. Results

About 45 animals were sectioned. The sharpness of the pictures is not very good, because they have three-dimensional information. Only stereo-colour slides show the whole information. Figs. 1—3 show different stages of the differentiating ear-placode. No special dissimilarities were seen between test and control animals. The strong variations could only be evaluated if many more experiments were made. An immediate influence of gravity on the structure and topographic position of the statolith membranes cannot be established from the present data. According to a NASA translation of a Russian paper, Vinnikov et al. [10] had a similar result from an experiment carried out in space. An extra finding not found in the literature was the laminated structure of the larval statoliths. This may allow a generalization of the hypothesis of Dohlman [9] about the formation of cupulae, i.e. that hairs from the sensory or sustaining cells regulate the formation of the extracellular structures in the vestibular organ. In our material one can see at least two systems of lamellae which interfere with each other. There is evidence for a steady re-building of lamellae if one compares the statoliths at different stages. This also supports the existence of functional hairs.

4. Discussion

How should we interpret these results in a phylogenetic sense and with respect to the development of function of the gravity sensors? One could make three speculations:

(1) Perception of gravity or acceleration is a "random" invention of the organisms, perhaps by modification of an unspecialized general sensibility of living tissues that cannot be recognized today. Ontogeny of the extracellular parts of gravity receptors does not interact with the environment, but has the same autonomous differentiation as other cellular organs.

(2) There exists an intracellular sensitivity for vector forces, such as the hair cells which control the first orientational movements of the embryo. On that basis, with the aid of gravity, topographic adjustment of the statoliths takes place.

(3) The first polarizing imprints for the adjustment of a vector receptor occurs after the "tumblerdoll" orientation of the fertilized egg.

The results reported suggest that only the first alternative is true because mechanisms required by (2) and (3) cannot act under conditions of simulated weightlessness. Experiments have been started in our laboratory to detect signs of function in the early embryologic hair cells of the statolith epithelium using electron microscopic preparation.

References

[1] W. Briegleb et al., in: Proc. XXIst Int. Astronautical Congr., Konstanz 1970, North-Holland Publ. Co., Amsterdam 1971 (p. 252).
[2] W. Briegleb, Naturwissenschaften **55**, 397 (1968).
[3] H. M. De Burlet, in: Handb. der vergleich. Anatomie der Wirbeltiere II, 2. Hälfte, Urban und Schwarzenberg, Berlin, Wien 1934 (p. 1293).
[4] H. Lüdtke und H. Schölzel, Zool. Jb. Physiol. **72**, 291 (1966).
[5] M. Igarashi, in: 2nd Symp. on Role of the Vestibular Organs in Space Exploration, NASA SP-115, Washington 1966 (p. 47).
[6] W. Briegleb et al., Microscopica Acta **72**, 147 (1972).
[7] F. Kopsch, Die Entwicklung des braunen Grasfrosches *Rana fusca* Roesel, Georg Thieme Verlag, Stuttgart 1952.
[8] W. Briegleb et al., in: Proc. XXth Int. Astronautical Congr., Mar del Plata 1969, Part 4: Bioastronautics, Pergamon Press, Oxford 1972 (p. 859).
[9] G. F. Dohlman, in: Vestibular Funtion on Earth and in Space, Pergamon Press, Oxford 1970 (p. 1).
[10] Ya. A. Vinnikov et al., Zh. Evol. Biokhim. Fiziol. (Leningrad) **8**, 343 (1972).

Life Science and Space Research XII — Akademie-Verlag, Berlin 1974

THE STABILIZING EFFECT ON THE TRUNK OF LABYRINTH AND NECK REFLEXES ACTING TOGETHER ON THE LIMBS

T. D. M. ROBERTS

Institute of Physiology, University of Glasgow, Glasgow, UK

The positional reflexes from the labyrinth upon the limbs have been re-examined in decerebrate cats with a technique involving denervations and clampings. The results have been used to design a model in which the effects on the limbs in response to the neck reflexes and to the positional reflexes from the labyrinth are each represented by mechanical linkages. Manipulation of the model reveals that when these two sets of reflexes are present together, the combined effect is to stabilize the trunk in relation to that direction in which it is appropriate to exert supporting forces in the prevailing field of accelerations.

If an animal's head moves while the trunk is stationary, some distortion of the neck is usually unavoidable. Consequently the attitude of the limbs may be influenced simultaneously by reflexes generated in the labyrinth and by reflexes generated in the intervertebral joints of the neck [1]. Some separation of neck reflexes from labyrinth reflexes has been achieved in the decerebrate cat by a combination of denervations and clampings [2]. The effects on the limbs in the positional reflexes from the labyrinth have been shown to be asymmetrical, in contrast to the symmetrical scheme outlined by Magnus [3] and hitherto generally accepted. The reflexes as now revealed can be described in the phrase "downhill limbs extend", with the corollary that the uphill limbs flex. This description applies to responses to tilting of the head about any horizontal axis.

Limb responses in the neck reflexes corresponding to lateral tilting of the head have been examined in the absence of labyrinth reflexes by twisting the axis vertebra while holding the head still, the first two intervertebral joints being denervated [4]. It has been confirmed that the direction of the responses corresponds to the phrase "chin limbs extend". The effects of interaction between the neck reflexes and the positional reflexes from the labyrinth have been studied with the aid of a model in which each reflex effect is simulated by mechanical linkages so adjusted that when the head is tilted while the trunk is stationary, the net change in the limbs is zero, in agreement with observations on intact animals.

The stimulus to the otolith organs has been taken as related to the direction and magnitude of the resultant of the contact forces acting on the skull on the following grounds [5]. The force of gravity itself cannot generate relative deformation within the otolith apparatus because, although the denser parts have more force acting on them, their resulting accelerations are the same as for the less dense parts. On the other hand, the forces that normally support the skull and prevent it falling to the ground are partly converted to shearing forces at the surface of the macula in consequence of the asymmetrical density of the otolith contents due to the presence of the otoconia. Forces producing horizontal accelerations have similar effects and it is therefore appropriate to combine all the contact forces to determine the direction and magnitude of the stimulus in the positional reflexes from the labyrinth.

By manipulating the model it becomes clear that the combination of the neck reflexes with the labyrinth reflexes acts on the limbs in such a way as to serve to stabilize the attitude of the trunk in relation to the appropriate direction for supporting forces, in spite of the fact that the directional sensor is in the skull and not in the trunk. The scheme represented by the model successfully predicts the attitudes of the limbs and trunk in various conditions such as standing on uneven groumd, standing in an aircraft during a banked turn, and so on, and it shows that in each attitude the head is free to move without reflex constraint.

References

[1] T. D. M. ROBERTS, Neurophysiology of Postural Mechanisms, Butterworth, London 1967.
[2] T. D. M. ROBERTS J. Physiol. **211**, 5 P (1970).
[3] R. MAGNUS and A. DE KLEIJN, Pflugers Arch. **145**, 455 (1912).
[4] K. W. LINDSAY, T. D. M. ROBERTS and J. ROSENBERG, in: Proc. 2nd Int. Symp. on Motor Control, Varna, Bulgaria, Oct. 1972 (1972).
[5] T. D. M. ROBERTS, Aerospace Med. **44**, 484 (1973).

Planetary Quarantine

Life Sciences and Space Research XII — Akademie-Verlag, Berlin 1974

TEN YEARS OF DEVELOPMENT OF THE PLANETARY QUARANTINE PROGRAM OF THE UNITED STATES

L. B. HALL

National Aeronautics and Space Administration, Washington, D.C., USA

A decade of comprehensive experimentation and research has provided a firm scientific data base for the formulation of justifiable standards and compliance measures for the United States Planetary Quarantine Program. This paper notes the significant achievements in quantitatively defining the quarantine problems and in developing usable values for key quarantine parameters describing biocidal effects of outer space, the release of micro-organisms from spacecraft reaching other planets and terrestrial microbial growth on other planets. Advances in compliance methods and techniques resulting from this research effort are also summarized with emphasis upon decontamination and sterilization, selecting of a standardized test organism and the assay of surface and buried contamination. As a result of this work, the United States can continue to assure fulfilment of international agreements for safeguarding the planets of our solar system with minimum burden placed upon space exploration programs.

1. Introduction

The United States Planetary Quarantine Program, administered by the National Aeronautics and Space Administration, has made significant strides in the past ten years in the establishment of realistic quarantine standards and in the development of means for verifying space flight project compliance with these standards. These tasks have not been easy ones. Factual knowledge needed for the quantification of justifiable specifications was often lacking. In its absence, some advocated adopting the "worst case", safe approach. This, however, was to be avoided since it could impose unnecessarily stringent requirements upon space flight projects and thereby jeopardize the scientific missions of the space exploration programs.

To solve these and other problems of planetary quarantine, a comprehensive experimentation and research program comprising more than 400 man years of effort has been pursued*. Many of the needed experimental data are now available and have been integrated into the program with the result that the United States has achieved a planetary quarantine solution which both safeguards the planets from biological infection and imposes minimum constraints upon flight projects.

* George Washington University, Washington, D.C., Dept. of Medical and Public Affairs, Biological Sci. Communication Project, Bibliography of Scientific Publications and Presentations relating to Planetary Quarantine, April 1973.

This solution is based upon risk management principles employing statistical probabilities whose values were derived through experimentation and research. The salient features of this research program are the subject of this paper.

2. Planetary Quarantine Basis

In response to a growing world-wide concern in the scientific community, the International Council of Scientific Unions (ICSU), at its 1958 meeting, founded the *ad hoc* Committee on Contamination by Extra-Terrestrial Exploration (CETEX). This group recommended a code of conduct aimed at achieving a compromise between all-out space exploration and absolute protection of the planets. It recommended that responsibility for overseeing planetary quarantine be turned over to the newly formed Committee on Space Research (COSPAR). COSPAR, in turn, accepted the responsibility and established a figure of 1×10^{-3} as a reasonable limit for the probability of contamination of a planet during the quarantine period. This action recognized an important and basic concept: that the probability of a foreign organism proliferating in a new environment is not necessarily unity. Thus a number of organisms can, if necessary in the interests of exploration, be landed on a planet if there is assurance that they will not be able to grow in that environment.

Further to define the limits of the problem, COSPAR estimated that the maximum number of missions that will be flown to Mars during the period of biological exploration would not exceed 50. In accord with these actions, the United States Planetary Quarantine Program has allocated an appropriate proportion of the risk to each planned mission and has required the mission managers so to design and operate the mission that the assigned risk will not be exceeded. Policy and detailed technical requirements are imposed upon each flight mission, inspections are carried out to assure compliance and the mission must be certified to be meeting the requirements before launch is authorized.

3. NASA Policy

In responding to the COSPAR requirements, the United States Planetary Quarantine Program has adopted a policy based upon Article IX of the United Nations Space Treaty, requiring the development and use of quantitative Planetary Quarantine standards. This policy is continuously reviewed and modified as indicated by the results of research and by the recommendations of pertinent groups such as the ICSU, COSPAR and the Space Science Board of the United States National Academy of Sciences. The American Institute of Biological Sciences has established a Planetary Quarantine Panel composed of microbiologists, mathematicians and engineers to advise the Planetary Quarantine Office in planning and evaluating research and quarantine performance. The membership of this Panel is shown in Table 1.

4. Planetary Quarantine Research

Originally the entire concept of Planetary Quarantine by CETEX and COSPAR was to sterilize, sterilize, sterilize! This was felt to be the only solution. To satisfy the purists on this matter, the spacecraft could be sterilized only by heating to incandescence, as one would a wire loop in the laboratory, but the spacecraft would then undoubtedly fail in its functions. To find a more practical solution, the United States entered into a research program designed to identify and

Table 1

American Institute of Biological Sciences Planetary Quarantine Panel

Present Members	Past Members
Prof. Richard G. Bond, Univ. of Minnesota, Chairman	Dr. Byron W. Brown, Jr., Stanford Univ. Medical Center
Dr. Robert Angelotti, Food and Drug Administration	Prof. Richard Cornell, Florida State Univ.
Dr. John Bagby, Jr., Colorado State Univ.	Dr. Frank B. Engley, Univ. of Missouri
Dr. John H. Brewer, Hardin-Simmons Univ.	Prof. Thomas W. Kethley, Georgia Institute of Technology
Mr. William B. Briggs, McDonnell Douglas Astronautics Co.	Dr. James C. Konen, Consultant; Ashland Chemicals
Dr. Allan H. Brown, Univ. of Pennsylvania	Prof. Irving J. Pflug, Univ. of Minnesota
Mr. Mark A. Chatigny, Univ. of California Berkeley	Dr. Gerald Silverman, MIT
Dr. Franklin A. Graybill, Colorado State Univ.	*Dr. Wolf V. Vishniac, Univ. of Rochester
Dr. Gilbert V. Levin, Biospherics, Inc.	Dr. William G. Walter, Montana State Univ.
Dr. Morton W. Miller, Univ. of Rochester	Mr. Robert P. Wolfson, Exotech Systems Inc.
Dr. Richard W. Porter, General Electric Co.	**Panel Consultant**
Dr. Orr E. Reynolds, American Physiology Society	Prof. Irving J. Pflug, Univ. of Minnesota
Mr. H. D. Sivinski, Sandia Corporation	**AIBS Coordinator**
Dr. H. Earle Swim, Univ. of Kentucky	M. F. Thompson
Dr. John Ulrich, Univ. of New Mexico	

quantify the effect of each of the factors influencing planetary contamination. Table 2 lists both the contributory factors and those that tend to prevent contamination of the planet. On the contributory side the bioload is of major importance. The bioload is the original source of the contamination—it may come from the open spaces on the spacecraft, from the covered surfaces, from the mated surfaces (that is organisms between two facing plates) or from organisms buried inside the solids.

Growth of organisms can occur on spacecraft materials that are nutritive. Micro-organisms can be ejected from the spacecraft in transit and find their way to the planet. There can be release of micro-organisms in the case of either an accidental crash landing or a nominal landing with subsequent erosion of the landed material. On the other hand, we have a number of factors that apply in the direction of preventing contamination. In the first place, decontamination of the spacecraft reduces the hazard. Sterilization reduces it even further—completely, of course, if sterilization is carried out to the true meaning of the word, complete absence of life. Fortunately, some of the materials in a spacecraft can produce autosterilization. Most notable among these are some of the fuels such as hydrazine. Production heating is a strong preventive factor since many spacecraft parts are heated to high temperatures for long periods of time during the manufacturing processes.

* Deceased

There appears to be a significant though minimal die-off of vegetative organisms during space transit due, perhaps, to age or dehydration. If temperatures are high enough there is a definite lethality from the combined stresses of space vacuum and temperature; however, the temperature of most spacecraft parts seldom reaches this level and this force is not counted upon at the present time.

Table 2
Contamination of a Planet

Contributory	Preventive
Bioload	Decontamination
Surface, open	Sterilization
Surface, covered	Autosterilization
Mated	Production heating
Encapsulated	Space transit-time
Growth on spacecraft	Space vacuum-temperature
Recontamination of lander	Solar wind
in launch operations	Trapped radiation
in earth atmosphere	Reliability of guidance
in flight	Entry heating
Accidental ejection	Non-release organism
Maneuver shocks	Lack of nutrients
Exhaust	
Micrometeoroid impact	
Accidental impact	
Microbial release	
Crash landing	
Erosion	

Ultraviolet radiation, even in secondary reflection, is a powerful sterilant. Many spacecraft, however, do not see the sun on all sides during their transit time and orientation has to be considered when evaluating the effect of ultraviolet radiation. Proton radiation may be of considerable importance particularly in the trapped radiation belts but there is a lack of hard information. We are investigating this factor further.

Reliability of guidance has proven to be a very dependable force in preventing contamination of the planet from a spacecraft intended to fly by or orbit the planet. With the degree of reliability currently available, there is very little chance of a spacecraft accidentally impacting a planet or satellite. Entry heating is an important factor in some of the missions depending on entry velocity and density of the atmosphere. The non-release of organisms results in only a small fraction of the organisms on or in the spacecraft being actually released to the planet surface. Finally, a lack of nutrients on the planet's surface or in its atmosphere tends to prevent growth of any organisms that may have reached the planet.

Over the years many methods of decontamination and sterilization have been examined. Some of these are listed in Table 3. Wet heat is out of consideration because of its damage to spacecraft materials, especially electronics parts. Ethylene oxide and its variants have been considered for use in United States Programs but have been rejected by the engineers because of the need for high humidity which would be damaging to the electronics and because of unfortunate experiences with the early use of ethylene oxide on the Ranger series of spacecraft. Brute force radiation is unacceptable for it is as damaging to spacecraft materials as is heat. However, thermoradiation, that is the use of low levels of gamma radiation

Table 3
Methods of Decontamination/Sterilization

Method	Completed	In progress
Wet heat	×	
Dry heat (target 30 June 1973)		×
Ethylene oxide (ETO)	×	
ETO plus methyl bromide	×	
ETO plus dimethyl sulfoxide	×	
Radiation	×	
Thermoradiation	×	
Plasma (target 30 June 1974)		×
Laser		

at moderate temperatures, offers promise. The exotic methods of plasma and laser sterilization are under investigation for possible application as surface sterilants on future spacecraft projects.

It must be pointed out that the majority of this research has, of necessity, been carried out with a standardized test organism, *Bacillus subtilis* var. *niger* because the organisms normally found on spacecraft cannot be collected in sufficient quantities to supply investigative needs. The United States Planetary Quarantine Program is now, however, engaged in validating its preliminary research results with naturally occurring organisms in the limited quantities that can be obtained.

Great steps have been made in these fields of research in the past ten years. We now examine some of the results in more detail.

5. Lethality of Outer Space

It was long suspected that mere exposure of micro-organisms to the sudden decompression of launch and the effects of the forces of outer space would be lethal. This proved not to be so. The launch pressure profile of a typical mission, shown in Fig. 1, shows the decompression experienced by micro-organisms on a spacecraft during launch. While some lethality as the result of the decompression was noted, as shown in Table 4, there is not sufficient consistency in the test results to warrant assigning a reduction term for this action. Similarly, organisms exposed for long periods of time to the combination of vacuum and temperatures simulating those of outer space show little, if any, die off. Although these tests show lethality at temperatures above 40 °C, such temperatures are not commonly experienced in flight and data are insufficient to warrant the assumption of a consistent reduction in the surface bioburden at this time.

Other forces in outer space, particularly those of trapped electrons and protons, may have pronounced effects on the survival of both vegetative cells and spores. Experiments by the Jet Propulsion Laboratory (JPL) of exposure to electrons under the test conditions shown in Table 5 resulted in appreciable death of the test organisms. The test results are summarized in Fig. 2. While the electrons at an energy level of 2 MeV did not produce sterilization, the number of both vegetative cells and spores were reduced by at least one decade and the effect

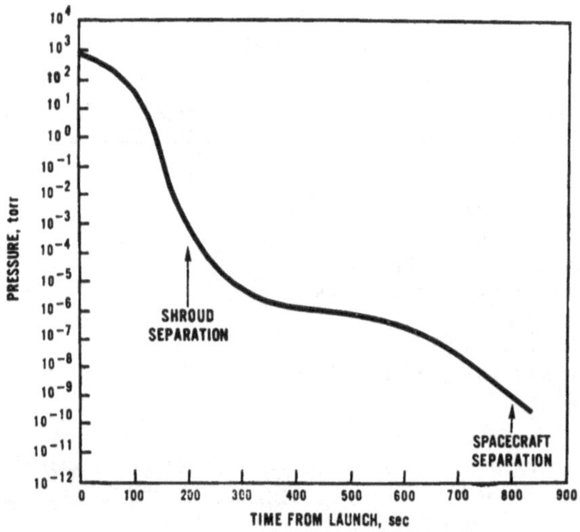

Fig. 1. Spacecraft freestream static pressure (*courtesy* Jet Propulsion Laboratory).

was uniform over the entire surface without regard to orientation. The effect of proton radiation has also been the subject of experimentation, but so far at only the 3-keV level. The evidence currently available at 3 keV indicates that approximately 50% of the exposed organisms will survive. The lethality of ultraviolet (UV) radiation has been the subject of extensive experimentation for medical and allied purposes and little, if any, further research is needed to define the results of UV exposure of spacecraft surfaces. Ultraviolet radiation lethality is a function of dose and duration of exposure. Cells (spores or vegetatives), if shown to be unshielded and exposed to UV radiation near the earth can be expected to die off in 1 hour to the extent of a 4 log reduction (i.e. to 10^{-4} of its value). Results must be carefully interpeted in terms of expected in-flight spacecraft orientation relative to the sun.

Table 4

Microbial Survival during Launch Phase

Type of micro-organisms	Test Duration/Temperature	% Reduction
Bacterial spores	12 min/20°C 1 hr/−20°C and 20°C 24 hr/−20°C and 20°C	0 to 44%
Non-sporeforming bacteria	12 min/20°C 1 hr/−20°C and 20°C 24 hr/−20°C and 20°C	0 to 60%

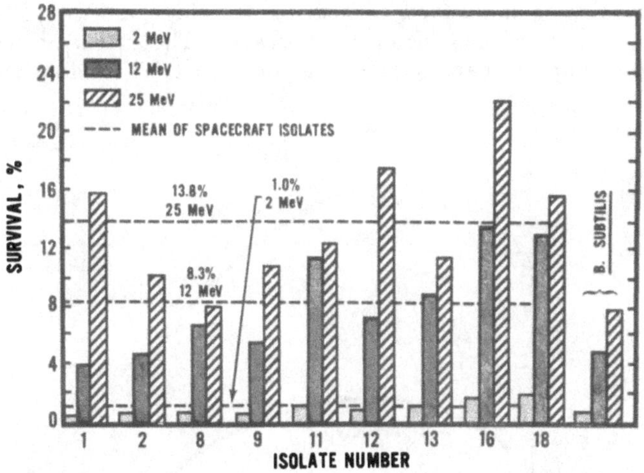

Fig. 2. Microbial survival in electron radiation (*courtesy* Jet Propulsion Laboratory).

Table 5
Test Conditions — Electron Radiation

Organisms	9 spores and 3 non-sporeformers isolated from Mars Mariner '71 *Bacillus subtilis* var *niger* *Staphylococcus epidermidis*
Dose	150, 300 and 450 kilorads
Temperature	—20°C and +20°C
Particle energy	2, 12 and 25 MeV
Flux	10^9, 10^{10} and 10^{11} e cm^{-2} s^{-1}
Pressure	10^{-6} torr

6. Probability of Microbial Release

Release of viable micro-organisms from the host spacecraft to the target planet's surface is requisite to planetary contamination. Organisms located on the exposed surfaces of a landed spacecraft can be expected to be released; however, many of those contained within enclosed sub-assemblies, located between mated surfaces or deeply embedded within materials will normally not be released to the planet environment. There are two mechanisms which are felt capable of releasing such non-exposed contamination from a Martian lander such as Viking. These are: (i) the break up of the spacecraft material resulting from accidental impact or high velocity crash landing, and (ii) aeolian erosion of the spacecraft material following landing.

Research has been undertaken to establish the degree of release of embedded contamination and the extent of attendant lethality for expected erosion and possible impact conditions. Experiments were conducted by the Boeing Company, under the guidance of JPL, in which pellets of spacecraft-type material were inoculated with a known contamination load and were fired at several different velocities to impact on targets of differing hardness. The resulting fragments were collected and biologically assayed. The extent of viable release as a function of impact characteristics was measured. These data are shown in Fig. 3.

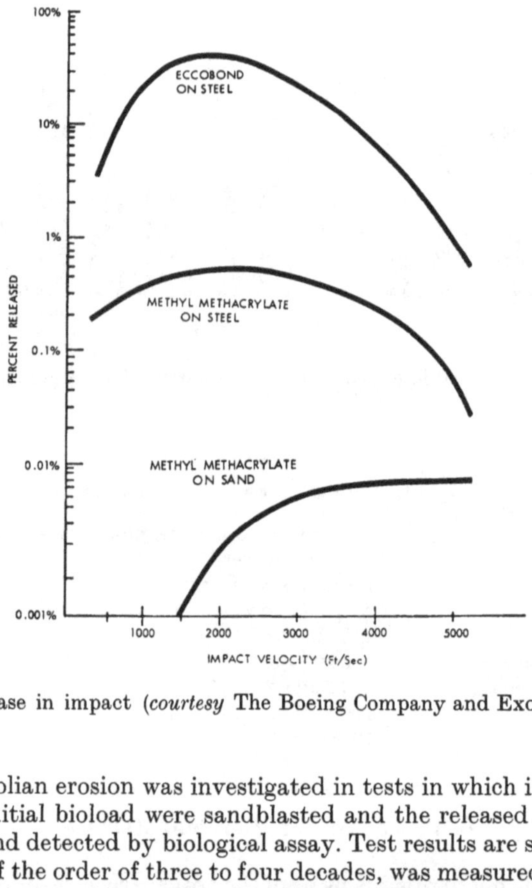

Fig. 3. Viable release in impact (*courtesy* The Boeing Company and Exotech Systems Inc.).

Release by aeolian erosion was investigated in tests in which inoculated pellets with a known initial bioload were sandblasted and the released micro-organisms were collected and detected by biological assay. Test results are shown in Table 6. Extensive kill, of the order of three to four decades, was measured. It is concluded that even though release by erosion is highly likely for spacecraft materials landed on Mars, viable release due to erosion will be very small because of the lethality of the erosion process.

Table 6

Erosion Release from Seeded Pellets: Experimental Results

Medium	Particle velocity (ft s^{-1})	Rate of erosion (m/yr)	Fraction of released survivors Methyl methacrylate	Eccobond
CO_2	300	30	1.1×10^{-3}	3×10^{-4}
Air	375	30	2×10^{-4}	2×10^{-5}
	334	7.5	1×10^{-4}	2×10^{-4}
	185	0.6	2×10^{-4}	1×10^{-4}

7. Probability of Growth (*P*(g))

The probability of growth of terrestrial organisms on foreign planets is basic to any consideration of planetary quarantine procedures and this factor has received considerable attention in our research program. The quantification of probability of growth for Mars was the subject of a meeting of a group of planetologists and microbiologists assembled by the Space Science Board of the National Academy of Sciences on 16 and 17 July 1970 at Woods Hole, Massachusetts. In defining environmental extremes under which terrestrial life forms could survive and grow, the group reviewed work conducted by the Ames Research Center. This Center assigned numerical limits to life-related factors including radiation, gas constituents, temperature, pressure and salinity. Based upon these results and those from other studies, the group developed the statement of minimal conditions for micro-environments capable of supporting the growth of terrestrial organisms as shown in Table 7. The presumed environmental conditions on Mars were then

Table 7

Definition of Micro-environment (conditions for growth of "hardy" terrestrial organism)

1. Water activity 0.95 or greater.
2. Temperature above 0° C for at least one-half hour per Martian day (for at least part of Martian year).
3. Nutrients available, including small amounts of water-soluble nitrogen, sulphur, phosphorus, and carbon (or CO_2 with light), and pH value between 5 and 8.
4. Attenuation of ultraviolet radiation by 1000 or more.
5. Absence of antimetabolites or "poisons".

reviewed in terms of this statement. An even/odds estimate of 1 chance in 100 was presented as the likelihood that defined minimum conditions for growth will exist on Mars near the Viking landing site. There appears to be little reason for modifying this minimum growth value at this time.

The next consideration in the group's review involved the fraction of delivered micro-organisms that would be hardy enough to grow in the micro-environment encountered. Here again, experimental results were applied in establishing a quantitative value. Tests conducted in the Ames Martian Simulator and a device at the IIT Research Institute provided growth rate data for such micro-organisms as: *Bacillus cereus* var. *mycoides*, *Interbacter aerogenes*, *B. subtilis* var. *niger*, and *Pseudomonas* (sp.). Fig. 4 presents a sample of these data.

Combining all of these considerations, the group developed estimates for the probability that a single randomly selected terrestrial micro-organism reaching Mars will grow and proliferate. These estimates as shown in Table 8 ranged from much less than 1 chance in 10^9 to a maximum of 1 chance in 10000. The even-odds estimate was stated as 3 chances in 10^9. Statisticians have attempted to apply confidence intervals to these estimates. If one assumes confidence levels of 0.5 for even-odds, and 0.999 for the upper limit, and assuming that the relationship between confidence level and probability estimate is log-normal (i.e. the logarithm of the confidence level value is a linear function of the probability estimate value), one can develop a basis for a value of the probability of growth *P*(g) of 10^{-6} with a confidence level of 0.96. This seems to be a reasonable level of confidence and this value, 10^{-6}, is being used for planning purposes by the United States Martian projects.

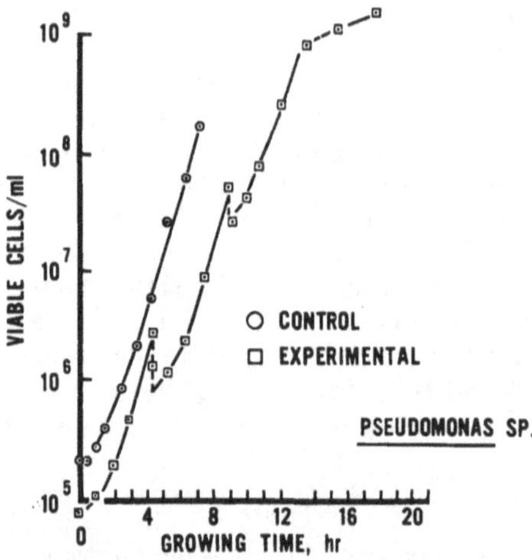

Fig. 4. Microbial growth under freeze-thaw conditions (*courtesy* Dr. R. Young, NASA Ames Research Center).

Table 8
NAS/SSB Reevaluation Sept. 1970

Condition	Even odds estimate	1000 to 1 upper limit
Likelihood that defined minimum conditions for growth will exist in micro-environments on Mars, near landing site	1/100	1
Fraction of total bioburden that could grow in micro-environment having defined minimum conditions	3/10000	1/100
Probability that released micro-organisms would be transported in viable condition (and adequate numbers) to micro-environment	1/1000*	1/100*
"Probability" that a released micro-organism would grow	3/1000000000	1/10000

* Includes "low-inoculum" factor of ∼ 1/10.

8. Research in Microbiological Assay

The assessment of microbial contamination levels, especially bacterial spores in and on space hardware, is one of the essential factors in the implementation and certification of spacecraft sterilization and decontamination. Sterilization needs are based, in part, on the number of bacteria present on the spacecraft, as determined by microbiological assays conducted during assembly. The success of decontamination (but not of sterilization) is measured by assays conducted on the completed spacecraft prior to launch. The precise enumeration of cells on any surface necessitates consideration of a number of physical problems. For example, many materials contain substances that inhibit the growth of micro-organisms and this inhibition must be neutralized. Assay procedures must be designed to separate the micro-organisms from the surfaces. Micro-organisms which have been present for long periods of time or have survived decontamination treatments, however, tend to adhere quite tenaciously to surfaces, making removal difficult. Also, micro-organisms in nature usually do not occur as single cells but rather in clumps and aggregates made up of many cells. These effects can markedly affect the accuracy of the assay. Several research groups, particularly the US

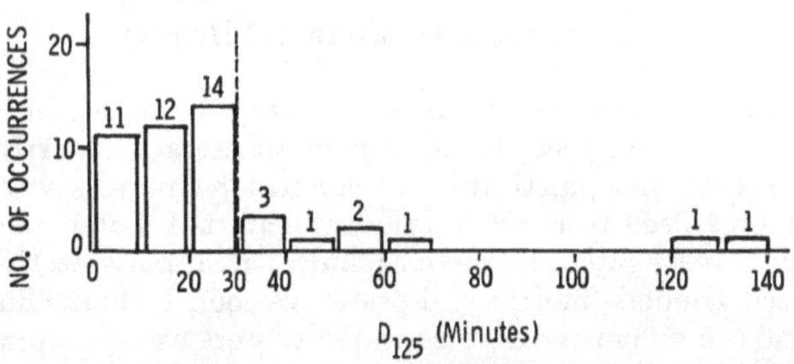

Fig. 5. D_{125} values from naturally occurring bacterial spores from soil samples from variou areas of the USA (from CDC Phoenix data).

Public Health Service and the Food and Drug Administration have studied these aspects of the problem and have developed standard assay techniques which have been applied to a number of spacecraft in programs such as Ranger, Lunar Orbiter, Surveyor, Apollo, IMP, Mariner, and Pioneer.

The need to control the microbiological burden on the spacecraft and the desire

Table 9

D_{125C} Values of Spore Isolates from Spacecraft (by CDC Phoenix)

Isolate	Source	Sporulation medium*	D_{125C} (min)	Isolate	Source	Sporulation medium*	D_{125C} (min)
CK-2	Surveyor 6	TAM	15	CK-12	Surveyor 6	AK 2	10
CK-2	Surveyor 6	AK 2	14	CK-13	Surveyor 6	AK 2	7.5
CK-4	Surveyor 6	TAM	46	CK-14	Surveyor 6	AK 2	7.5
CK-4	Surveyor 6	AK 2	19	CK-17	Surveyor 6	AK 2	3.6
CK-5	Surveyor 6	TAM	30	CK-19	Apollo 20	AK 2	15
CK-5	Surveyor 6	AK 2	20	CK-21	Apollo 20	AK 2	2.5
CK-6	Surveyor 6	AK 2	8	CK-22	Apollo 20	AK 2	11
CK-8	Surveyor 6	AK 2	10	CK-23	Apollo 20	AK 2	27
CK-9	Surveyor 6	AK 2	5	CK-24	Apollo 20	AK 2	1.4
CK-10	Surveyor 6	AK 2	10				

* Both media were supplemented with 80 ppm calcium chloride and 20 ppm magnesium sulfate.

to reduce the severity of the final sterilization cycle has prompted the establishment of various contamination control techniques, such as use of clean rooms, filtration and personal barriers during manufacture and assembly of spacecraft. NASA-sponsored research in these areas has made significant contributions to the technology.

Further to verify the sterilization cycle, the inactivation characteristics of naturally occurring organisms have been, and are being, studied. These organisms were obtained from locations representing sites of manufacture, assembly and launch of various United States spacecraft programs. Fig. 5 shows the range of D_{125} values for these bacteria. Typical D values for spore isolates obtained from actual spacecraft are indicated by the data in Table 9.

9. Encapsulated Microbial Density

Research has been undertaken to develop a basis for estimating the number of viable organisms that may be buried within spacecraft material. In the early 1960s several rather unsophisticated mechanical techniques were used for recovering micro-organisms from the interiors of materials. Later, chemical methods of recovery were investigated. In the mechanical area, pulverization of materials, use of high speed grinders, mortar and pestle, various drill attachments, ball and vortex mills and other abrasive devices and saws were used in attempts to recover viable organisms from the materials. Most of these techniques were fraught with difficulties in that the yields from the materials were very hard to correlate. Particle size was virtually impossible to control and the death of organisms that occurred through these processes left a large question as to the validity of the data. Subsequently, a biogrinder was developed at NASA's Marshall Space Flight Center which produced the controllable and consistent particle size needed for assay on non-metallic material. Results of the biogrinder experiments with actual electronic piece parts representative of the Mariner Mars spacecraft are illustrated in Table 10*. Here we see the assay items to be resistors, semiconductors of various types, capacitors (both wrapped and foil types), and wire-wound resistors. The last item in this list represents non-electronic piece parts in coated and uncoated circuit boards. Piece parts, which comprise some 70% of the non-metallic elements of a spacecraft, contain from 3 to 150 organisms per cubic centimeter. For a typical spacecraft 130 organisms per cubic centimeter of non-metallic material can be considered as a conservative value based upon these experimental data. These few examples serve to illustrate the efficacy of the United States experimental program in eliminating information gaps and facilitating the development of quantitative planetary quarantine requirements which can be technically substantiated and effectively utilized.

Research data have been employed in an analytical methodology providing an orderly and systematic way to examine the various, significant, potentially contaminating sources and to estimate their probabilities of producing contamination. This examination includes a means of trading off the multitude of interrelated considerations so that the results and associated costs can be justified. As pertinent knowledge has increased, due, in part, to sponsored research, it has

* Results are also shown for assays performed by disassembling piece parts that could be handled in this manner.

Table 10
Examples from Survey of Experimental Data

Investigator/ Organization	Assay item	No. of items	Total volume V_E (cm³)	Fracturing procedure	Est. f	Est. P_{SP}	No. of positive ΣX_i
Petersen & Favero PHS/Phoenix	Resistor fixed composition 0.855 cm³	40	34	Broken in half exposed 40 cm²	1.2	1	0
(1970–1971)	Semiconductor rectifier	10	14	Broken in half exposed 19.5 cm²	1.4	1	0
	Semiconductor rectifier	30	42	Pulverized	10^4	5×10^{-1}	1
	Coil, Adjustable core (7.8 g; 4.0 cm³)	40	160	Pulverized	10^4	5×10^{-1}	0
	Resistor fixed wire-wound (4.8 g; 2.5 cm³)	27	68.5	Disassembled	10^4	1	3
	Capacitor, mylar/aluminum foil (3.5 g; 2.2 cm³)	27	59.4	Disassembled	10^5	1	1
	Resistors		5.65	Biogrinder	10^5	10^{-1}–10^2	32
	Hot-molded capacitors		7.2	Biogrinder	10^5	10^{-1}–10^2	3
	Semiconductor rectifiers		3.44	Biogrinder	10^5	10^{-1}–10^2	9
	Coated/uncoated circuit boards		10.10	Biogrinder	10^5	10^{-1}–10^2	2

become possible to relax the constraints for compliance with contamination control requirements. The point in our analysis of planetary quarantine has now been reached where sterilization of all the releasable microbial burden is no longer essential. A finite, but small, number of terrestrial micro-organisms can, in some cases, be permitted to reach the planetary surface without violating the basic contamination requirement. Decontamination techniques (as opposed to sterilization) can now be considered as appropriate control measures for many space flight missions.

Future effort in United States planetary quarantine research will be directed toward those parameters whose values are uncertain. Many of the "PQ" parameters applicable to the outer planets fall into this category. The continuation of the research approach described herein is expected to benefit future flight projects and continue to provide assurance of the compliance of the United States Planetary Program with international planetary quarantine agreements.

Life Sciences and Space Research XII — Akademie-Verlag, Berlin 1974

VERIFICATION OF THE EFFICACY OF SPACECRAFT STERILIZATION

V. I. Vashkov, N. V. Ramkova, G. V. Scheglova, L. Z. Skala and A. G. Nekhorosheva

Moscow, USSR

The possibility that terrestrial micro-organisms can survive decontamination and be ejected into the environment of planets emphasizes the need for control of sterilization, and the development of standards based on models made of relevant materials, which can be carefully investigated. Control of sterilization can be by physical, chemical or biological (bacteriological) means, depending on the purpose; bacteriological methods are the most precise and most generally useful.

Test objects should be artificially infected by cultures of micro-organisms which are highly resistant to the sterilizing agent being tested and which are of high stability. To ensure reliable sterilization the density of contamination of test objects should be 10—100 times the natural level of contamination. The nature of the surface on which the test micro-organism is placed affects the efficacy of sterilization owing to differences in heat transmission or sorption of chemical disinfectants. The role of the recovery medium is also important.

Physical and chemical methods are particularly valuable as indicators of efficacy of sterilization, e.g. the use of chemical indicators of temperature in heat sterilization.

Results by various methods of control are in good agreement, but improvements in techniques are still necessary to make spacecraft sterilization more reliable.

Terrestrial micro-organisms can sometimes survive decontamination or final sterilization processes and after reaching a planet these surviving micro-organisms can be ejected into the environment of the planet. The importance of considering the probability of the ejection of microbes was pointed out by Sagan, Levinthal and Lederberg [1] and by Horowitz, Sharp and Davies [2] when discussing the problem of contamination of Mars. Thus great attention is being given all over the world to the control of sterilization [3—10]. It is considered that the efficacy of automatic space vehicles' sterilization must be controlled at the stage of development of recommendations on sterilization by appropriate methods of the spacecraft as a whole or of its components. Final judgment on the efficacy of the chosen method, on the correctness of the technology and the organization of the sterilization process should be based on sterilization of models which can be dismantled as necessary and carefully investigated.

There are three different methods for control of the efficacy of sterilization: physical, chemical and biological (bacteriological). Choice of the method depends on the sterilization techniques and on the concrete aims of control. The bacteriological method is the most specific and precise and is suitable for all types of sterilization techniques. The other methods (physical and chemical) can also contribute, and can be used for the operative control of sterilization processes—

they estimate the quantitative parameters of sterilizing agents, e.g. temperature, amount of sterilizing gas and integral dose of ionizing radiation.

During the development of sterilization methods the most important control is verification of the effectiveness of sterilization using artificial contamination. We consider it expedient not to control sterilization of the spacecraft materials and components before launch, during the period when spacecraft sterilization is carried out, so that the spacecraft is not exposed to the danger of repeated contamination, but instead we prefer to follow precisely the recommendations on sterilization that have been worked out beforehand.

Despite the existing work on the development of sterilization control techniques, it is still necessary to investigate and improve certain aspects of the problem. In our opinion attention must be paid to determination of the standard stability of test micro-organisms, in order to choose the most appropriate ones as test objects. One must measure the recovery after sterilization and choose the best media (or chemical indicators) for the control of the sterilization process. The efficacy of sterilization depends, among other things, on the resistance of the micro-organisms contaminating the component under study. Therefore in order to provide a safety margin micro-organisms that are highly resistant to the sterilizing agent must be used. At present there are no official standards for the resistance of cultures. We do not consider it expedient to regulate the species of the test micro-organisms; even when the conditions of control are standardized, the resistance of the test cultures to the given sterilizing agent must be precisely determined.

As a rule, micro-rganisms highly resistant to the main sterilizing agents are found among sporeforming saprofites like *Bacillus subtilis*, *B. mesentericus*, inhabiting various environments, but mostly in soils rich in organic matter—so-called garden soil. We have tested about 2000 samples of garden soil and studied 111 strains of sporeforming micro-organisms like *B. subtilis* and *B. mesentericus* that were resistant to sterilizing gases, heat and ionizing radiation, and obtained 82 strains of sporeforming cultures from these samples. After estimation of resistance of these strains to the sterilizing agents we selected some test-strains, the resistance of which can be considered as standard. Thus, for the control of the efficacy of thermal (air) sterilization, test cultures that survived 100° for 3 hours could be used. For control of gas sterilization we use cultures that can survive 2.5 hours exposure to mixture "OB" (ethylene oxide with methyl bromide in the proportion 1:1.44 w/w) at 35 °C and in concentration 1000 mg/l. For control of radiation sterilization we use cultures that survive when 1 to 2×10^9 cells are suspended in water after 2.5 megarad of radiation.

The nature of the surface on which the test micro-organism is fixed influences the efficacy of sterilization; this can be explained by differences in heat transmission or sorption of chemical disinfectants. When sterilized by hot air, micro-organisms on metal plates are killed more quiclky than micro-organisms in soil. On the contrary, when sterilized by "OB" the microflora in soil is killed more quickly than on metal plates or plastic surfaces, due to the greater adsorbing capacity of the soil. Evidently soil samples should not be used for control of sterilization methods, even more because the microbial floras of different soils, or different samples of the same soil, are often quite dissimilar. We believe that samples of garden soil should be used only as a test for steam sterilization. For control of other methods and for developing sterilization schedules, it is necessary to use test objects made of the same material as the object to be sterilized. The test objects should be artificially infected by cultures of micro-organisms highly resistant to the given sterilizing agent. They should be placed in many different parts of space vehicles and their components.

The extent of microbial contamination of the object and the suitability of the recovery medium are also of considerable importance. Efficacy of the "cold sterilization" methods (radiation and gas) depends a great deal on the extent of contamination of the object. A dose of 2.0 megarad is ineffective if the contamination is over 10^6 microbes per square centimetre; with 2.5 megarad the corresponding value is 10^8. Microbial burden affects the results of gas sterilization in a similar way. Thus to ensure reliable sterilization the density of contamination of the test objects used for control of sterilization should be 10—100 times higher than the natural contamination of the object studied. In addition, the natural microbial burden of spacecraft and their components should be kept low by using aseptic and antiseptic methods of assembling under conditions of low contamination. It is especially important to reduce the burden when sterilization is by "cold" methods.

To determine the role of the recovery medium the suitability of the following media was studied: thioglycolate broth, lucerne broth, meat-peptone broth, Hottinger broth, and casein broth made from tryptic or acid digests. The lucerne medium gave the highest recoveries with *Bacillus subtilis* exposed to ionizing radiation (0.5—1 megarad): the recoveries were about 2 times higher than with other media. *Streptococcus faecium* grows equally well in lucerne medium, tryptic casein broth and Hottinger broth. The most favourable media for *Bacillus subtilis* exposed to the "OB" mixture (1 g per 1 l of the mixture, 180 min exposure) are the lucerne broth, thioglycolate broth, tryptic casein broth and Hottinger broth. Meat-peptone broth is not so suitable. Recovery of viable *Bacillus subtilis* spores is greatest in thioglycolate and lucerne broths after exposure to dry heat (160° for 45 min). In casein broth the recoveries are 10—15 times less.

For cultures exposed to saturated steam the lucerne and thioglycolate broths and the medium made from pancreatic digest of casein were found to give highest recoveries. Broth made from acid-hydrolysed casein is least suitable. For organisms exposed to chemical sterilizing agents (iodonte, peracetic acid, hydrogen peroxide) the thioglycolate medium gave the lowest recoveries. The other media were all about the same in this respect.

Physical and chemical means of control serve as indirect indicators of the efficasy of sterilization. Heat is one of the main methods of spacecraft sterilization, so selection of chemical indicators that record the time when the maximum temperature is reached is of considerable interest. For this purpose pure chemicals melting at different temperatures can be used. Examples with their melting pointc are: benzoic acid, 119°; iodoform, 119°; succinimide, 126°; succinic acid anhydride, 131°; salicylic acid, 155°; succinic acid, 185°; dinitrophenylhydrazine, 195°. One can also use thermoindicator dyes that change their colour as a function not only of temperature but also of the time of exposure to high temperature. Such dyes are being developed which can show that the temperature has been kept at 160° for 60 min, 170 °C for 45—60 min and 180 °C for 30—40 min. When appropriate indicators are used simultaneously with test micro-organisms, the results of biological and physical controls are in good agreement, but we still need improvements in these techniques to make spacecraft sterilization more reliable.

References

[1] C. Sagan, E. C. Levinthal and J. Lederberg, Science, 1191 (1968).
[2] N. H. Horowitz, R. P. Sharp and R. W. Davies, Science, **155**, 1501 (1967).
[3] E. Steiger and W. Suckov, in: Medizin und Pharmazie, Heft 2 (1967).
[4] M. C. Reynolds, The feasibility of thermoradiation for sterilization of spacecraft, a preliminary report, 1969.
[5] M. Alexander, Quarantine for samples from Mars, Cornell University, New York.
[6] A. K. Astafieva, N. V. Ramkova, L. Z. Skala, L. M. Vorontsova, L. P. Mazurova, Yu. P. Volkov and I. P. Strelets, in: Proc. Int. Symp. on Disinfection and Sterilization 1972 (p. 63).
[7] I. P. Dishkant, Transactions of CNIDI, **19**, 203 (1970).
[8] N. V. Ramkova and I. P. Dishkant, Transactions of VNIIDIS, **21**, 167 (1971).
[9] N. V. Ramkova, Transactions of CNIDI, **20**, 150 (1969).
[10] M. A. Tumanjan, Z. G. Pershina, T. V. Golosova, Ya. N. Rashban and L. A. Nikolaeva, in: Proc. VIII Conference on Problems of radiation immunology and microbiology, Moscow 1972 (p. 86).

Life Sciences and Space Research XII — Akademie-Verlag, Berlin 1974

LUNAR SAMPLE QUARANTINE PROCEDURES: INTERACTION WITH NON-QUARANTINE EXPERIMENTS

M. B. DUKE and M. A. REYNOLDS

NASA, Lyndon B. Johnson Space Center, Houston, Texas, USA

The success of NASA in conducting a joint program of biomedical quarantine experiments and a wide range of scientific experiments on the lunar samples is well known. Some of the problems, especially those where the quarantine requirements interacted with non-quarantine experiments, should not be forgotten. Vacuum systems constructed for sample handling proved inadequate; stainless steel biological containment cabinets purged with dry nitrogen at slightly lower than atmospheric pressure, with autoclaves or sterilizers on entrances or exits, were more satisfactory, but leaks introduced atmospheric contamination. Sterilization of sample handling systems and sample transfer containers introduced organic contaminants. Analytical equipment built into glove box systems was clumsy and less sensitive than necessary. Sample requirements for quarantine testing competed with those for physical and chemical experiments. Many of the undesirable interactions could be magnified if a planetary mission with much smaller sample return is undertaken. However, experience gained in the lunar quarantine should provide basic information in the design of planetary sample quarantine facilities and procedures that can minimize the hazards and maximize the scientific return.

1. Introduction

In 1964, the National Academy of Sciences Conference on Back Contamination recommended that crew, equipment and materials returned from the moon or planets be subject to quarantine, as an important part of a set of procedures designed to provide the maximum feasible protection of the earth's biosphere from potential contamination by lunar pathogens [1]. The Conference also recognized the need to provide the greatest feasible protection of the samples from degradation that would reduce their scientific value, a primary interest of the scientific community. Quarantine and an elaborate scheme of biomedical testing were carried out for the Apollo 11, 12 and 14 missions and, although the quarantine restrictions were suspended for the Apollo 15, 16 and 17 missions, biomedical testing continued with samples from those missions.

Systems and procedures [2, 3] established to conduct quarantine operations in the Lunar Receiving Laboratory (LRL) in some cases conflicted with the interests of the physical sciences community in investigating and preserving samples for scientific investigation. The conflicts arose in four basic areas: (i) containment requirements, (ii) sterilization procedures, (iii) system design for early analysis, and (iv) competition for sample materials. This paper briefly describes some of the conflicts; many were satisfactorily solved, others were not, but all can serve as illustrative examples for future considerations of quarantine procedures.

2. Containment

Containment of potentially harmful agents was the principal goal of the quarantine program and efforts were made at all stages to minimize the potential spread of contamination [2]. Containers were designed to seal samples under vacuum on the moon, "housekeeping" procedures were designed to remove contamination during the flight back, recovery procedures attempted to minimize exposure of lunar material to the earth, and the LRL was operated at the "state-of-the-art" of biological containment. The most serious conflicts between biological containment and sample contamination occurred in the controlled environment systems in which samples were opened, studied and prepared for analysis in the LRL. Biological containment systems used in working with terrestrial pathogens served as models for much of the LRL design. These consisted of stainless steel glove boxes purged with high purity nitrogen gas at less than atmospheric pressure (negative pressure) so that leaks would only be inward; gas was exhausted through a low vacuum system passed through an incinerator for sterilization [3]. Nitrogen was chosen, because it was expected and subsequently demonstrated that lunar materials were altered by exposure to oxygen and water. Unfortunately, the best cabinets leak, and gases diffuse through gloves, so control of water vapor and oxygen in the cabinet atmosphere was poor during negative pressure operation. After relief of the quarantine restriction, the nitrogen cabinets were operated at greater than atmospheric pressure (positive pressure) and significant reductions of oxygen and water vapor contaminants resulted. Comparative levels were: H_2O, negative pressure 85—125 ppm, positive pressure 15—25 ppm; O_2, negative pressure 25—50 ppm, positive pressure 10—30 ppm. With negative pressure operations, occasional small leaks caused both O_2 and H_2O to rise as high as 2000 ppm. Under positive pressure conditions, observable large leaks (holes in gloves) produced no detectable rise in gaseous contaminants.

Prior to Apollo 11, a complex gloved vacuum chamber system was developed for opening the vacuum sealed sample containers and examining the samples [3]. The vacuum system was preferred by some elements of the scientific community, because the samples had been in a high vacuum on the moon and their potential reactivity to gases was unknown. This system, which was designed to meet quarantine requirements, proved to be vulnerable to sample contamination from occasional massive leaks (failures) and continually from organic contaminants in the pumping systems.

None of the LRL systems satisfactorily eliminated atmospheric contamination and, thereby, some scientific data were lost. For example, the D/H ratio in the solar wind cannot be adequately determined in the lunar samples because they have been contaminated by small quantities of terrestrial water [4]. Glove box systems are probably inadequate for complete contamination control, and should perhaps be replaced by remote-controlled manipulation systems in future missions. Design of a successful, reliable and adequately redundant remote system would require a major development effort, probably comparable with the effort of developing the LRL vacuum system.

3. Sterilization

Guidelines for sterilization were designed to be the most stringent normally used for the most resistant terrestrial infectious organisms, namely spores [2]. A variety of sterilization techniques and sterilants were used at various steps in the sample handling, all of which represented some hazard to the samples. Table 1 lists the principal sterilization techniques and their potential hazard to the samples [5]. Sterilization of the vacuum system and nitrogen cabinets before the mission was not specified by the containment requirement, but was considered good biological practice to insure that no terrestrial organisms would contaminate the samples and be mistaken for lunar organisms in the quarantine tests [3]. Heat sterilization was used in the vacuum systems and presented no contamination hazard. In the nitrogen systems, a variety of liquid and gaseous sterilants was tried. The initial attempt to sterilize the major nitrogen processing line using formalin failed, with complete fouling of the plumbing systems, and the cabinets were rendered inoperable. Later, sterilization using ethylene oxide (ETO) gas was demonstrated to be satisfactory from a biological point of view; however, residual organic contamination resulted from polymerization of the ETO [6]. Ottawa sand, a natural quartz sand, which was used as a monitor to collect the organic constituents for analysis, was itself a contamination hazard to the quartz-free lunar samples.

Table 1
Sterilizing Agents for Lunar Sample Operations

Sterilant	Results	Contaminants of samples resulting from use
Heat	Sample and container sterilized	None
Sodium hyperchlorite (NaOCl)	Surface only	Strong oxidizing agent, chlorine, sodium
Peracetic acid (CH_3CO_2OH)	Surface only	Hydrogen peroxide, acetic acid, 2,6-pyridine dicarboxylic acid
Formaldehyde (HCHO)	Surface only	Polymers of formaldehyde
Ethylene oxide (CH_2CH_2O)	Surface only	Dioctyl phthalate, hydrocarbons, polymers of ethylene oxide

Pre-mission sterilization, in retrospect, appears to have been unnecessary. In Apollo 15, 16 and 17, the nitrogen cabinets were not sterilized prior to the mission, but were rigorously cleaned [7]. The microbiological tests made on samples from those missions have shown no evidence of contamination by terrestrial microorganisms [8], indicating that cleaning and prolonged exposure to dry nitrogen gas was sufficient to reduce contamination by terrestrial organisms to levels that were undetectable with the analytical procedures that were used. Early acceptance of thorough cleaning in lieu sterilization would have eliminated many of the organic and inorganic contamination problems experienced during the quarantine missions (the complexity and cost of cabinet systems also would have been greatly reduced). Sterilization also was necessary whenever a sample was removed from one biologically isolated system to another. If a container was verified leak-tight, it could be sterilized on the surface by soaking it in one of several solutions (Table 1); however, residual sterilant was a contamination hazard to the samples. Several types of resealable containers were developed prior to Apollo 11, which provided first-level sample protection. Reliable apparatus and procedures for heat sealing samples in teflon bags were developed following Apollo 12. Multiple sealing of samples and careful handling procedures then provided ample protection from contamination by sterilants.

In cases where samples required sterilization so that they could be removed from the primary barrier for examination or analysis, heat sterilization was used [3, 5]. Temperature/time relations derived to reach "sterile" conditions ranged from 700 hours at 100 °C to 3 minutes at 190 °C. These sterilizations could be performed without hazard of organic or inorganic contamination, but at the risk of destruction of thermally controlled phenomena. For example, it has been found that solar wind or meteoritic bombardment of the lunar surface apparently mobilizes certain volatile elements (e.g. Hg, Pb, Ar), allowing them to escape or move at temperatures as low as those used in heat sterilization, see, e.g. [9]. For purposes of the lunar sample preliminary examination, only small portions of samples were sterilized and transferred for analysis (chemical analysis, organic and noble gas mass spectrometry, optical microscopy, etc.). For planetary sample return missions, the possibility of heat sterilizing the entire sample should be carefully assessed, both from the point of view of minimizing quarantine requirements and minimizing organic and inorganic contamination.

4. Analytical System Design

Prior to Apollo 11, the concern of the biomedical community with respect to containment was amplified by the concern of the physical scientists that the samples could be chemically highly reactive. The incorporation of analytical equipment into controlled environment sample handling facilities was one of the responses to these concerns. An emission spectrograph, X-ray diffraction and X-ray spectroscopy equipment (following Apollo 11), and optical microscopy equipment were incorporated into the nitrogen cabinetry in a configuration that would permit samples to be transferred for analysis without leaving the controlled environment [3]. These "built-in" systems had two principal disadvantages: (i) they were not as effective (i.e. less sensitive, more difficult to operate) as analytical equipment operating outside the controlled environment cabinets and (ii) they were complicated systems and used materials that could contaminate samples; thus, facilities were designed that could separate the analytical laboratories from the cabinets in which the bulk of the samples were stored and studied. The sample transfer procedures for moving materials between the sample handling cabinets and the analytical instruments were more complex than for transfer to an outside laboratory. For example, a pneumatic system from the main vacuum chamber to the analytical laboratory was constructed (but never used).

A different approach was used in the design of the Radiation Counting Laboratory (gamma-spectrometry), which was constructed sixty feet below ground and a considerable distance from the vacuum chamber. Samples for the RCL were transferred in sealed containers which were sterilized by immersion. The containers were not opened in the RCL, and the samples could be adequately protected by proper packaging. Whereas the chemical and X-ray analysis done with "built-in" equipment was not comparable in precision or sensitivity with similar outside laboratories, the RCL operated as a highly competitive analytical laboratory.

On return of the first lunar samples, it became immediately apparent that the samples were not highly reactive. It had also become clear that samples were much more easily manipulated and transferred to analysis if the analytical equipment was outside the controlled environment cabinetry, and the emphasis turned from operation of the installed systems to sterilization of samples and analysis in separated laboratories. From Apollo 14 onward the controlled environment operations were limited to relatively simple operations, including (i) weighing, (ii) photography, (iii) binocular microscope examination (microscopes outside cabinets), and (iv) splitting and packaging of samples for transfer to the analytical laboratories. All analytical work was performed outside the controlled environment facilities.

5. Sample Competition

Competition for samples caused conflicts between the physical and biomedical science groups. Many chemical and isotopic analyses require only tens to hundreds of milligrams of sample to make determinations of the chemical composition, mineralogical relationships and age of the samples, so that emphasis could be placed on the analysis of highly selected portions of the lunar samples, such as small rock fragments from the soil. In comparison, the quarantine biomedical testing, from which negative scientific results were anticipated and resulted, required major quantities of lunar material representing "average" properties of the samples. The approximate amounts used for biomedical testing on the six Apollo missions were: Apollo 11, 700 g; Apollo 12, 528 g; Apollo 14, 589 g; Apollo 15, 144 g; Apollo 16, 73 g; Apollo 17, 61 g. The preliminary examination experiments (physical science experiments) had no specified minimum requirement, but the largest sample use occurred in Apollo 17, where approximately 65 grams was used, of which two-thirds is retained as encapsulated material for preparation of microscope sections. Of course, much larger quantities were distributed for detailed scientific study following the preliminary examination.

The quantities of material used in the lunar quarantine testing are probably large in relation to the return capabilities of unmanned planetary missions (the Russian Luna 16 mission returned 101 grams [10]). The conflicts between biomedical and physical sciences experiments may be even greater for planetary missions than for the lunar program, but could be greatly reduced by reducing the amount of sample material required for the experiments. Several proposals can be offered for dealing with the conflict; for example, if a planetary quarantine program is undertaken:

(i) *Joint use of samples.* The requirements of both classes of experiments can be met with less sample consumption if careful selection of materials is made. Deep drill samples from Apollo 15—17 were studied jointly, with the emphasis of sample use placed on finer grained constituents by the biomedical tests, and coarser grained constituents by the physical sciences experiments (constituents finer than $^{1}/_{10}$ mm make up more than half of every lunar soil).

(ii) *Recovery of samples.* Experiments can be designed to recover useful material following tests. For example, mild leaching of samples with organic solvents leaves most lunar minerals unaltered for mineralogical study.

(iii) *Miniaturization of experiments.* Great advances have been made in analytical instrumentation in the physical sciences. In many cases, the prospect of having limited lunar sample materials forced the development of techniques which used samples one or two orders of magnitude smaller than were required before Apollo. Definition of new procedures which are more sensitive and attention to miniaturization may be important aspects of the development of protocols for biological testing of returned planetary samples, if sample quantities are small.

6. Conclusions

Many of the conflicts between physical and biomedical sciences groups during the lunar quarantine program could have been solved prior to Apollo, if adequate resources and time had been available to develop and test the systems and their interactions. A decision is necessary on the question of sample quarantine requirements; if samples are sterilized, quarantine requirements can be reduced and sample contamination minimized. However, if samples are to be quarantined, and biomedical testing is to be carried out, early decisions are necessary and a great deal of work will be necessary to design appropriate facilities and procedures, so that the potential scientific accomplishments of planetary sample-return missions are not jeopardized.

References

[1] A. H. BROWN, Conf. on Potential Hazards of Back Contamination from the Planets, NASA/NRC (1964).
[2] B. C. WOOLEY, NASA Tech. Note NASA TN D-6856 (1972).
[3] J. C. MCLANE, JR., et al., Science **155**, 525 (1967).
[4] S. EPSTEIN and H. P. TAYLOR, JR., Geochim. Cosmochim. Acta Suppl. **2**, 1421 (1971).
[5] B. C. WOOLEY, Manned Spacecraft Center Document MSC-00018, Revision B (1970).
[6] D. A. FLORY and B. R. SIMONEIT, Space Life Sci. **3**, 457 (1972).
[7] P. GRAF, Manned Spacecraft Center Document, MSC-03243 (1971).
[8] G. R. TAYLOR and B. C. WOOLEY, in: Abstracts of papers presented at 4th Lunar Science Conf. 1973 (p. 711).
[9] S. JOVANOVICH and G. W. REED, JR., in: Abstracts of papers presented at 4th Lunar Science Conf. 1973 (p. 418).
[10] A. P. VINOGRADOV, Geochim. Cosmochim. Acta, Suppl. 2, 1 (1971).

Life Sciences and Space Research XII — Akademie-Verlag, Berlin 1974

VIABILITY OF *Bacillus subtilis* SPORES EXPOSED TO SPACE ENVIRONMENT IN THE M-191 EXPERIMENT SYSTEM ABOARD APOLLO 16

H. BÜCKER[a], G. HORNECK[a], H. WOLLENHAUPT[a], M. SCHWAGER[a] and G. R. TAYLOR[b]

[a] Arbeitsgruppe für biophysikalische Weltraumforschung, Universität Frankfurt, Frankfurt/Main, FRG
[b] NASA Johnson Space Center, Houston, Texas, USA

During the Apollo 16 space flight, in the experiment system M-191, (microbial response to space environment) spores of *Bacillus subtilis* 168 were exposed to space vacuum or solar UV irradiation with a peak wavelength of 254 mm or both. The effects of these space factors on the colony-forming ability of the spores were studied. It was found (i) that space vacuum alone did not affect the survival of pre-dried spores; (ii) that space vacuum in combination with solar UV irradiation with a peak wavelength of 254 nm had a synergistic effect, which may by attributed to a UV supersensitivity of the spores during vacuum exposure. These results agreed with findings of simulation experiments on earth. It was concluded that air dried spores may survive exposure to space vacuum if shielded against solar UV irradiation.

1. Introduction

Bacillus subtilis spores have proved to be suitable for viability tests in simulated space environment. They are very resistant to vacuum exposure. Even in a 10^{-10} torr ultra-high vacuum about 90% of spores maintained their colony forming ability as compared with air dried spores [1—3]. Spore survival *in vacuo* may be severely influenced by a variety of different vacuum-chamber components such as pumping systems, chamber material, and gauges. Unwanted toxic effects, which could lead to incorrect interpretation of the biological response to space vacuum, may be present in the simulation chambers. Therefore biological studies in space are also necessary. Another interesting parameter of space flight is the solar UV radiation, which is known to be biologically important [4—6]. The strong damaging effect of solar UV on micro-organisms in space is in agreement with data from simulation experiments on earth, in showing supersensitivity of bacteria to ultraviolet when irradiated in a high vacuum [7—9]. This may be due to the production of photoproducts in the DNA that are not easily repaired, similar to the mechanism which has been demonstrated in cells of *Escherichia coli* [10].

The survival of *Bacillus subtilis* spores after exposure to space vacuum and solar UV irradiation is reported in this paper and compared with simulation experiments on earth. The data obtained strengthen the strict requirement for sterilization of planetary mission spacecraft.

2. Experiment Performance

Pure spore suspensions of *Bacillus subtilis* 168 were obtained by the following method: After incubation on Bacto-TAM sporulation agar (Difco) at 30 °C for 20 days, cells were re-suspended in distilled water, trated with lysozyme (0.75 mg ml^{-1}) for four hours and washed nine times in distilled water. Both liquid and dry samples were used in the experiment. Liquid samples were kept in Kel-F plastic cuvettes (about 50 μl capacity) with quartz windows [11], whereas the dry samples were prepared on glass supports (Fig. 1). About 2×10^5 spores per support

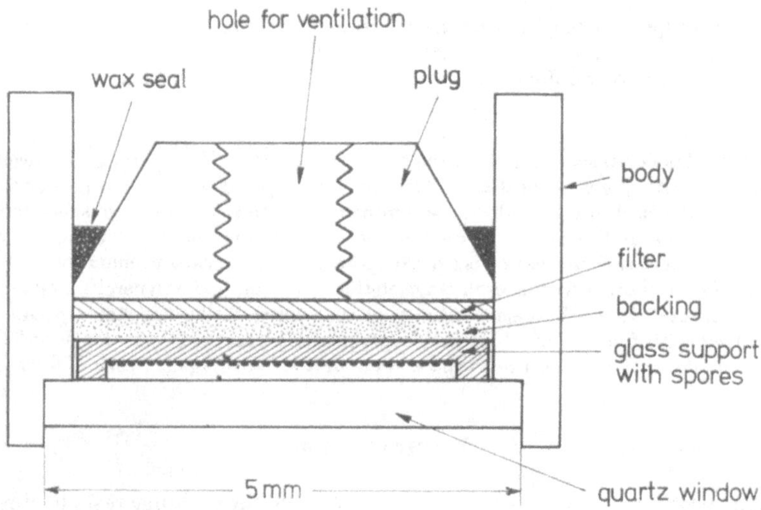

Fig. 1. Cuvette containing dry samples of *Bacillus subtilis* spores.

were dried for 24 hours over silica gel at 4 °C. The spores were spread in single layers to prevent protection by being overlaid.

The Microbial Ecology Evaluation Device (MEED), the flight hardware of the experiment M-191 aboard Apollo 16, contained 34 *Bacillus subtilis* 168 cuvettes. During flight, the samples were exposed to space vacuum for 1.3 hours and to one of three intensity levels of solar UV irradiation at a peak wavelength of 254 nm for 10 minutes. This exposure was accomplished during the extra-vehicular activity conducted by the command module pilot [12]. Data on the ultraviolet irradiation dose were obtained from evaluation of film actinometers and ferrioxalate actinometry solutions [12]. In addition, non-exposed flight controls were used. Further ground controls, launch vibration controls and a simultaneous laboratory simulation experiment were employed.

For post-flight analysis the liquid as well as the dry samples were transferred into phosphate-buffered saline. Vigorous shaking for five minutes resulted in suspension of the spores in the diluent. Appropriate dilutions were spread on Bacto nutrient broth agar (Difco). After incubation for about 20 hours at 37 °C the macrocolonies were counted. The survival was evaluated in terms of colony-forming ability compared with that of the controls.

3. Results and Discussion

During the Apollo 16 flight, the spores met with several random space flight factors, like launch vibration, weightlessness and variations in cabin atmosphere and pressure. The flight controls (12 flight cuvettes that were not specially treated) were analysed to determine the biological effect of these flight parameters. The survival of these flight controls did not differ significantly from the survival of the ground controls, either for liquid samples (Table 1) or for dry samples (Table 2).

Table 1

Effect of Random Space Flight Factors on the Viability of Spores of *Bacillus subtilis*

Sample	Treatment	Suspension in Distilled Water c.f./cuvette	Survival (%)	Suspension in PO_4 Saline c.f./cuvette	Survival (%)
Flight (Apollo 16)	Random space flight factors	$(1.89 \pm 0.38) \times 10^6$	100 ± 28.0	$(3.21 \pm 0.11) \times 10^6$	107 ± 7.4
Ground (Houston)	Vibration	$(2.32 \pm 0.15) \times 10^6$	120 ± 15.0	$(3.83 \pm 0.07) \times 10^6$	128 ± 5.8
	Control	$(1.89 \pm 0.16) \times 10^6$	100 ± 8.5	$(3.00 \pm 0.12) \times 10^6$	100 ± 4.0

c.f. = colony formers

It is known from the Biostack experiment [13] that the ionizing cosmic background radiation reaching the spores inside the spacecraft did not produce a statistically significant inactivation. The remaining 22 flight cuvettes were used for studies on the effects of space vacuum and solar UV irradiation of 254 nm, applied singly or in combination. After exposure to space vacuum for 1.3 hours, about 100% survival was observed on pre-dried flight samples (Table 2). Thus, space vacuum alone did not cause a detectable inactivation of air dried spores. Irradiation of liquid spore samples with solar UV (254 nm) resulted in inactivation values fitting well to the dose survival curve of the simulation test (Fig. 2). This shows that space flight conditions did not alter the UV sensitivity of the spores. When the spores were exposed to space vacuum and solar UV simultaneously, an increased

Table 2

Effect of Space Vacuum on the Viability of Dried Spores of *Bacillus subtilis*, prepared from a Phosphate-Saline Suspension

Sample	Treatment	c.f./cuvette	Survival (%)
Flight (Apollo 16)	Space vacuum (1.3 hrs)	$(1.19 \pm 0.04) \times 10^5$	103 ± 10.2
	Control	$(1.08 \pm 0.03) \times 10^5$	94 ± 9.7
Ground (Houston)	Vibration	$(1.22 \pm 0.08) \times 10^5$	106 ± 13.4
	Control	$(1.15 \pm 0.08) \times 10^5$	100 ± 6.9

c.f. = colony formers

UV sensitivity was observed (Fig. 2). This increase was comparable with that observed in simulation tests. It is assumed that the reduced water content of the spores in the vacuum environment of space favours the production of either fewer, or less reparable, DNA photoproducts. These products may cause the UV

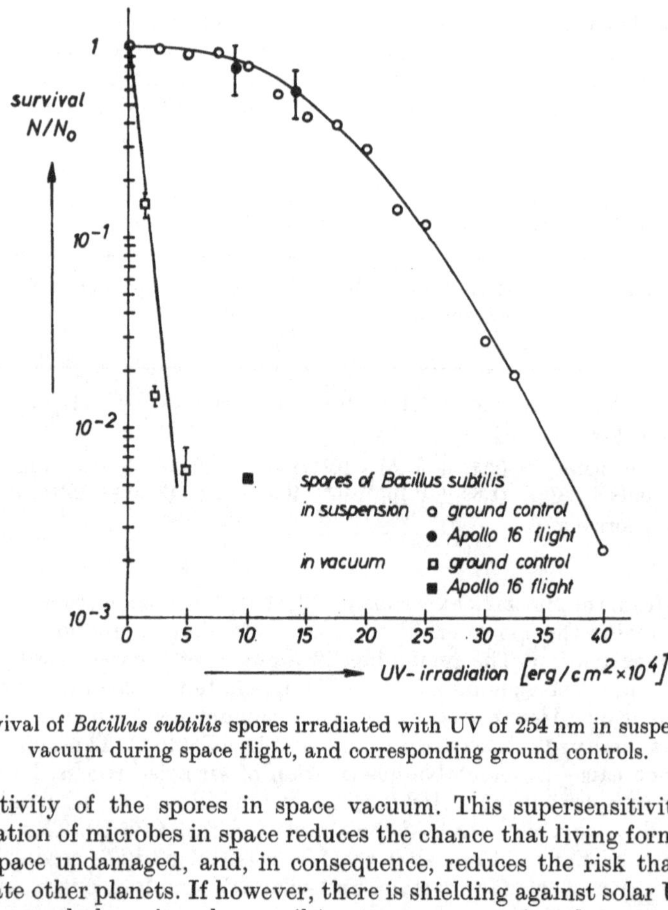

Fig. 2. Survival of *Bacillus subtilis* spores irradiated with UV of 254 nm in suspension and in vacuum during space flight, and corresponding ground controls.

supersensitivity of the spores in space vacuum. This supersensitivity to solar UV irradiation of microbes in space reduces the chance that living forms will pass through space undamaged, and, in consequence, reduces the risk that they will contaminate other planets. If however, there is shielding against solar UV irradiation (for example by minerals or soils) spores may survive the exposure to the harsh environment of space. Spacccraft sterilization is therefore necessary to prevent the transport of terrestrial viable micro-organisms to other planets.

References

[1] D. M. PORTNER, D. R. SPINER, R. K. HOFFMAN and C. R. PHILIPS, Science **134**, 2047 (1961).

[2] H. BÜCKER and G. HORNECK, paper presented at 8th Ann. Meeting of Deutsche Gesellschaft f. Biophysik, Giessen 1970.

[3] C. A. HAGEN, J. F. GODFREY and R. H. GREEN, Space Life Sci. **3**, 108 (1971).

[4] P. R. LORENZ, C. L. HEMENWAY and J. HOTCHIN, Life Sciences and Space Research VI, 100 (1968).

[5] J. Hotchin, P. Lorenz and C. L. Hemenway, Life Sciences and Space Research VI, 108 (1968).
[6] G. B. Orlob, Canad. J. Microbiol. **15**, 245 (1969).
[7] G. Horneck, H. Bücker and H. Wollenhaupt, Life Sciences and Space Research IX, 119 (1971).
[8] G. Horneck and H. Bücker, Strahlentherapie **141**, 732 (1971).
[9] H. Bücker, G. Horneck, Gr. Turcu and C. Thomas, Arch. Roum. Path. Exp. Microbiol. **31**, 7 (1972).
[10] M. Schwager, C. Thomas, G. Horneck and H. Bücker, paper presented at 6th Int. Congr. on Photobiology, Bochum 1972.
[11] C. E. Chassay and G. R. Taylor, in: Proc. Symp. on Microbial Response to Space Environment, NASA Tech. Memo. X-58103, 1973 (p. 21).
[12] G. R. Taylor, C. E. Chassay, W. L. Ellis, B. G. Foster, P. A. Volz, J. Spizizen, H. Bücker, R. T. Wrenn, R. C. Simmonds, R. A. Long, M. P. Parson, E. V. Benton, J. V. Bailey, B. C. Wooley and A. M. Heimpel, in: Apollo 16 Prelim. Sci. Rep., NASA SP-315, 1972 (p. 27-11).
[13] H. Bücker, G. Horneck, E. Reinholz, W. Scheuermann, W. Rüther, E. H. Graul, H. Planel, J. P. Soleilhavoup, P. Cüer, R. Kaiser, J. P. Massué, R. Pfohl, R. Schmitt, W. Enge, K. P. Bartholomä, R. Beaujean, K. Fukui, O. C. Allkofer, W. Heinrich, H. François, G. Portal, H. Kühn, H. Wollenhaupt and G. H. Bowman, in: Apollo 16 Prelim. Sci. Rep. NASA SP-315, 1972 (p. 27-1).

Life Sciences and Space Research XII — Akademie-Verlag, Berlin 1974

TERRESTRIAL QUARANTINE CONSIDERATIONS FOR UNMANNED SAMPLE RETURN MISSIONS

A. R. Hoffman, W. Stavro, L. W. Miller and D. M. Taylor

Jet Propulsion Laboratory, California Institute of Technology, Pasadena, Calif., USA

Numerous studies have shown that unmanned extraterrestrial sample return missions are technically feasible, with Mars the planet of primary interest. One consideration which has not received much attention in these studies is the possibility of contamination of the terrestrial environment by alien organisms returned with a vehicle containing an atmospheric or soil sample. For the purpose of understanding some of the possible implications of a terrestrial quarantine constraint on a mission and for developing a basic approach which can be used to demonstrate compliance beyond that developed for Apollo, a terrestrial quarantine study has been performed at the Jet Propulsion Laboratory. It is shown that some of the basic tools developed and used by the planetary quarantine community have applicability to terrestrial quarantine analysis. By using these tools, it is concluded that (i) the method of biasing the earth aiming point when returning from the planet is necessary but, by itself, may not satisfy terrestrial quarantine constraints; and (ii) spacecraft and container design significantly influence contamination transfer.

1. Introduction

The National Aeronautics and Space Administration (NASA) has established requirements for the biological quarantine of planets involved in unmannea exploration. All NASA unmanned planetary flight projects are governed by a uniform set of planetary quarantine (PQ) requirements as set forth in NHB 8020.12, "Planetary Quarantine Provisions for Unmanned Planetary Missions", April 1969. These planetary quarantine requirements do not apply to earth itself.

The present paper considers the quarantine aspects of returning samples to earth in unmanned planetary spacecraft. Some of these quarantine considerations have been studied extensively in the United States in conjunction with the manned lunar missions by the Interagency Committee on Back Contamination [1, 2]. Terrestrial quarantine (or back contamination) is defined as contamination of the terrestrial biosphere with organisms or materials of extraterrestrial origin. Unmanned planetary sample return missions have quarantine aspects which differ considerably from both manned and "one-way" unmanned missions.

2. Mission and Spacecraft Description

2.1. Mission Description

The phases of an unmanned sample return mission are given in Fig. 1. To provide familiarity with sample return concepts, a return mission from Mars was selected for this study. It is assumed that the outbound flight and the landing are analogous to that of a Mars lander such as described in [3]. The recovery and quarantine phase is assumed to be similar to the manned lunar missions described in [1, 2]. For the return phase it is assumed that an ascent vehicle containing

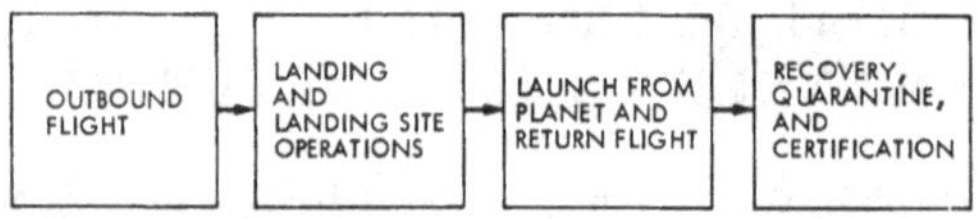

Fig. 1. Phases of unmanned sample return missions.

the Mars soil sample will rendezvous with a return vehicle in Mars orbit. The return vehicle is then injected from Mars orbit on a transfer trajectory to earth. The return flight of the mission can be appropriately described by considering the trajectory modes available as options and comparing the total mission durations, injection energies from Mars, and earth arrival velocities. On the return portion of the flight, the interplanetary trajectory can be of a conjunction or an opposition class (Fig. 2). The conjunction class is characterized by a minimum energy Mars-earth trajectory (i.e. minimum fuel requirements). In order to achieve this, the spacecraft must stay on Mars, or in Mars orbit, for a duration of over a year to await the desired planetary geometry. Conversely, the opposition class is character-

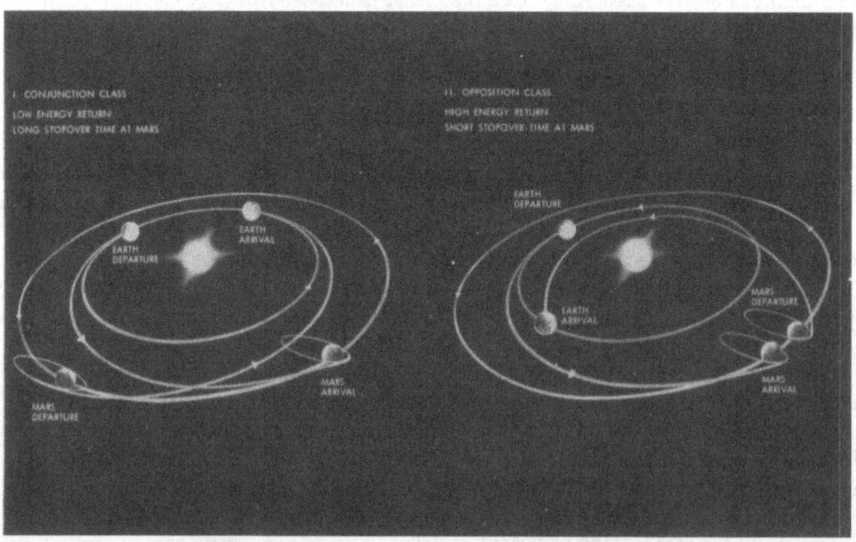

Fig. 2. Mars sample return trajectory options.

ized by a short stopover at Mars and shorter mission duration. However, the mission is penalized with a larger fuel requirement.

For this study the flight time to Mars is assumed to be approximately 200 days. The conjunction class return will result in a 509-day stay at Mars and an injection

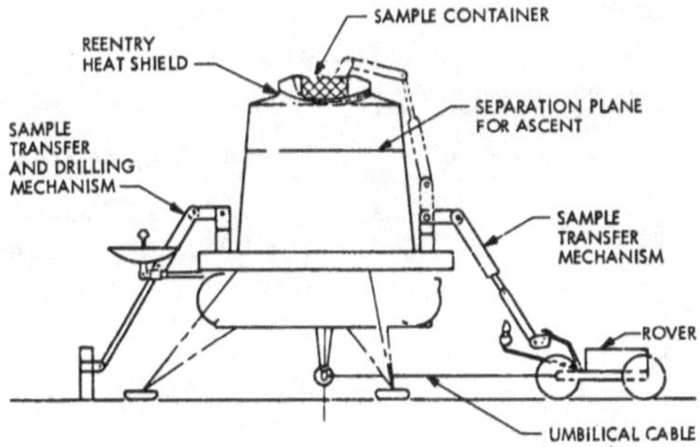

Fig. 3. Typical Mars lander/ascent vehicle.

energy from Mars orbit of 5.4 $km^2 s^{-2}$ with a resulting earth arrival velocity of 5.6 $km s^{-1}$. The total mission duration is approximately 1020 days. The opposition class return was determined for a 40-day stay at Mars, which results in an injection energy from Mars orbit of 48 $km^2 s^{-2}$, an earth arrival velocity of 14.3 $km s^{-1}$, and a total mission duration of 465 days. In summary, the conjunction class has *longer* mission duration, but *smaller* injection energy from Mars and *smaller* earth arrival velocity than the opposition class.

2.2. Spacecraft Description

A study by Langley Research Center depicted a set of typical spacecraft that could be used for a Mars sample return mission [4]. An example of a Mars descent lander from this study is shown in Fig. 3. In this design an indexable motor drive

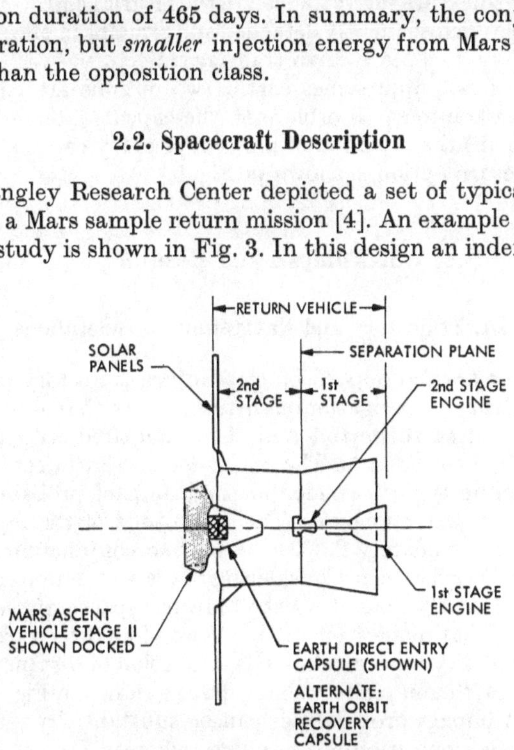

Fig. 4. Mars ascent vehicle shown mated with return vehicle.

sample container is on the top part of the vehicle and is mounted on the earth re-entry heat shield. The container receives its samples from a rover vehicle by means of the sample transfer arm. At liftoff, the ascent vehicle separates from the rest of the lander and will later rendezvous with the orbiting return vehicle. In Fig. 4 the ascent vehicle is shown mated with the return vehicle. The sample

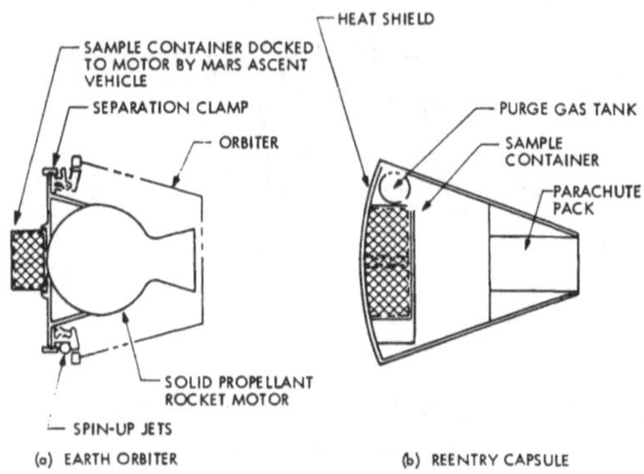

Fig. 5. Typical design for an earth orbiter and earth re-entry capsule.

container and heat shield are then attached to the return vehicle, and the remainder of the ascent vehicle is subsequently detached and remains in Mars orbit. The return vehicle then injects into a Mars-earth transfer trajectory.

As the return vehicle approaches earth, two options are considered: (i) the injection of a capsule into earth orbit with the capsule later being picked up by a space shuttle, or (ii) the injection of a direct re-entry capsule. A typical design for each sample return option is shown in Fig. 5.

3. Analysis and Results

3.1. Trajectory and Navigation Considerations

The conjunction class and opposition class return trajectory modes were examined in detail in terms of navigation maneuver errors that would result in probabilities of accidental earth impact and then compared with possible levels of terrestrial quarantine constraints. The out-of-Mars orbit injection maneuver was analyzed to determine typical accidental earth impact probabilities that would result from errors in that maneuver. The magnitude of the injection maneuver was found to be approximately 2.3 km s^{-1} for the conjunction class return and 7 km s^{-1} for the opposition return. Considering that an execution error in magnitude is usually about 0.1 of 1% (based on the Mariner spacecraft series experience), an error estimate in that maneuver of 1 m s^{-1} was found to be reasonable.

The impact probability was calculated as a function of distance from the center of earth (B) in the earth aim plane; the results are shown in Fig. 6. An interesting observation is that impact probabilities can be substantially reduced by biasing (increasing B) in the opposition class return, whereas for conjunction returns the impact probability is relatively unchanged by biasing. This was due to the fact that the error ellipse for the opposition class is more elliptical than for the conjunction class.

In summary, the results show that accidental earth impact probabilities due to likely execution errors in the out-of-Mars orbit injection maneuver will very likely

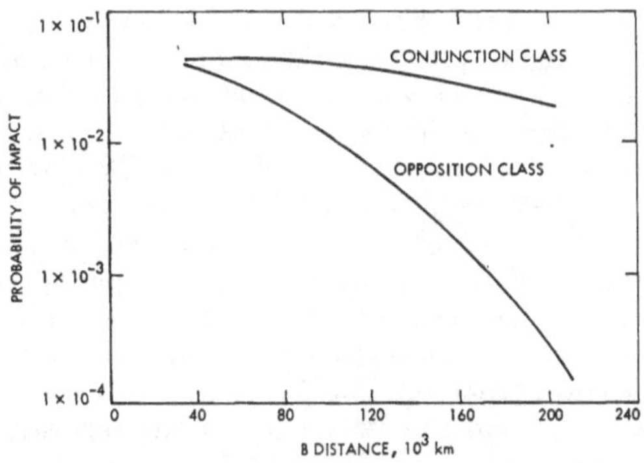

Fig. 6. Probability of earth impact by biasing Mars departure aim point.

be larger than 10^{-2} for reasonably desired aim points at earth. If this value is modified by the probability of not being able to perform a corrective maneuver if on an impact trajectory, typically 10^{-2}, then such an error would satisfy a terrestrial quarantine constraint of 10^{-4}. This assumes that the probability of earth contamination given accidental impact ($P_{C/I}$) is unity. (A value significantly less than unity would make the quarantine constraint easier to meet.)

3.2. Spacecraft Considerations

A typical Mars lander/ascent vehicle was shown in Fig. 3. As the vehicle resides on the surface, it is reasonable to assume that Martian material could transfer to the spacecraft surfaces by various natural or onboard mechanisms. Dust particles could become detached from the surface of the ascent vehicle and then be transferred to the orbiter surfaces during rendezvous by means of dynamic events such as vibration, contact of docking surfaces, and docking malfunctions.

The possibility of contamination transfer to the orbiting return vehicle is best discussed with the aid of Fig. 4. The direct re-entry capsule configuration (Fig. 4 with details in Fig. 5b) is used as an example for discussion of the contamination transfer mechanisms. When the ascent vehicle docks with the return vehicle, it is actually docking with the orbiter portion of the re-entry capsule. When the ascent vehicle disengages and falls away from the return vehicle, the heat shield and sample container remain attached to the orbiter portion of the re-entry capsule. The contamination on the sample container would be trapped in the interior of the re-entry capsule. Special seals and bioshields would be necessary to control possible problems at the conjunction of the heat shield and the rear portion. To summarize, spacecraft leaving the Mars surface will be contaminated. The contamination transfer problems from the ascent vehicle to the return vehicle require detailed study. Spacecraft design will make a considerable difference in magnitude of these problems.

Design attention should be directed to the sample container and the onboard monitoring necessary to assure integrity of the container. Specifically, an adequate seal for the sample container is required not only to prevent terrestrial contamination during the return flight and recovery, but also to preserve the integrity of the sample. In addition, the possible disastrous consequences of the release of sample contents into the earth's atmosphere due to meteorite impact or spacecraft breakup must be precluded. This could be accomplished by enclosing the sample container in a sphere with an outer ablative shield covering.

To determine the condition of the spacecraft as well as the integrity of the sample container during the landing phase of the mission, a TV system could be used to scan the surface of the ascent vehicle to (i) determine surface contamination from planet soil by comparing reflective properties of an uncontaminated control surface against exposed surfaces, and (ii) verify the proper operation of the sample gathering and encapsulation mechanisms. As the spacecraft returns to earth, strain gauges or leak detection sensors could monitor the condition of the sample container to detect leaks or break-up. The complexity of these design considerations will be strongly dependent on the stringency of the terrestrial quarantine constraints.

4. Conclusions

From the results of the trajectory and navigation analysis for a Mars-earth sample return trajectory, it is concluded that if a stringent terrestrial quarantine requirement is imposed, the constraint could not be satisfied by simply biasing the earth aim point. Specifically, investigation of the out-of-Mars orbit maneuver shows that a standard magnitude error in that maneuver would result in an earth impact probability of the order of 10^{-2}. Assuming a return midcourse maneuver "unreliability" of 10^{-2}, the overall probability of uncontrolled accidental impact would be 10^{-4}. Even if one assumes that accidental impact would result in the certainty of biological contamination, an overall terrestrial quarantine constraint of the order of 10^{-4} could still be satisfied. However, a constraint of the order of 10^{-8} would generally mean that terrestrial quarantine could not be satisfied by navigation strategy alone, i.e. biasing the earth aim point. Other methods of controlling and monitoring contamination transfer, such as spacecraft design, would become increasingly important.

Finally, it is concluded that if planning of unmanned planetary return missions is to proceed, a terrestrial quarantine policy should be established.

References

[1] A. H. BROWN, Conf. on Potential Hazards of Back Contamination from the Planets, Nat. Acad. Sciences, Nat. Res. Council, Space Science Board, Washington, D.C., February 1965.

[2] G. B. PHILLIPS, in: Planetary Quarantine, Gordon and Breach Science Publ., New York 1971 (p. 121).

[3] J. S. MARTIN and C. W. SIBERS, Project Viking, paper presented at 23rd Congr. of Int. Astronaut. Fed., Vienna, October 1972.

[4] A Study of Automated Mars Surface Sample Return Missions, NASA Langley Research Center, Hampton, Va, November 1970.

Life Science and Space Research XII — Akademie-Verlag, Berlin 1974

REDUCTION IN MICROBIAL BURDEN OF SPACECRAFT DUE TO HEATING ON ENTRY INTO THE ATMOSPHERE OF JUPITER

C. C. GONZALEZ, W. JAWORSKI, A. D. MCRONALD and A. R. HOFFMAN

Jet Propulsion Laboratory, California Institute of Technology, Pasadena, Calif., USA

Planetary quarantine analyses performed for recent unmanned Mars and Venus missions assumed that the probability of contamination by a spacecraft given accidental impact was equivalent to one. However, in the case of the gaseous outer planets, the heat generated during the inadvertent entry of a spacecraft into the planetary atmosphere might be sufficient to cause significant microbial burden reduction. This could affect navigation strategy by reducing the necessity for biasing the aim point away from the planets. An effort has been underway at Jet Propulsion Laboratory to develop the tools necessary to predict temperature histories for a typical spacecraft during inadvertent entry. In order that the results have general applicability, parametric analyses are performed. The thermal response of the spacecraft components and debris resulting from disintegration is determined. The temperature histories of small particles and composite materials, such as thermal blankets and an antenna, are given special attention. Guidelines are given to indicate the types of components and debris most likely to contain viable organisms, which could contaminate the lower layers of the Jovian atmosphere (~1 atmosphere pressure).

1. Introduction

With missions to the planets, an unmanned spacecraft has a finite probability of accidentally impacting the planet due to inherent errors and uncertainties in the spacecraft navigation system [1]. In performing planetary quarantine analysis for Mars missions, it has been assumed that impact of the planet by an unsterilized spacecraft would lead to a probability of contamination equal to one. However, due to the relatively large inertial entry speed of 61 km s^{-1} for Jupiter, the heat received by a body entering the planet's atmosphere will be significant. Thus, the reduction in microbial burden due to entry heating may be so large that the probability of contamination may be considerably less than one. Therefore a study was initiated to determine if the spacecraft and spacecraft debris receive sufficient thermal heating to result in a significant reduction of microbial burden.

2. Technical Discussion

2.1. Background

An artist's conception of a typical Jupiter flyby spacecraft used in this study is shown in Fig. 1. The spacecraft consists of a nine-sided frame assembly (1.5 m across by 0.5 m high), to which nine electronic chassis are attached. The frame assembly provides primary structural attachments for a high-gain antenna (3.7 m in diameter), propulsion assembly, radioisotope thermoelectric generator, science instruments, science platform, celestial sensors, attitude control assembly, and thermal blankets. A body entering the atmosphere of a planet first encounters

Fig. 1. Typical outer planet spacecraft.

the rarefied portion of the atmosphere, called the "free-molecular regime". In this regime, the energy transfer to the body results from exchanges of kinetic energy of colliding gas molecules with the body. The body then enters the denser regime of the atmosphere where the gas behaves like a homogeneous, continuous matter; this regime is referred to as the "continuum". Continuum flow regime entries are characterized by a strong high-density shock ahead of the body which results in the body surface becoming exposed to very high convective and radiative heat fluxes.

Accidental biological contamination of the atmosphere of Jupiter by an encountering spacecraft can arise from three sources: (i) the spacecraft itself, (ii) ejecta released during flight from earth, and (iii) disintegration debris (i.e. fragments and particles) released as the spacecraft passes through the atmosphere. The general survivability of these three sources has previously been determined for Jovian entry assuming simple configurations (spheres, circular disks, and cylinders) and homogeneity in the configuration material [2, 3]. This investigation concentrates on determining the thermal responses of selected components of the spacecraft, consisting of complex geometries and several different materials, and ejecta and disintegration debris as they pass through the free-molecular flow regime and early continuum.

2.2. Method of Analysis

Classical thermodynamic equations were used to determine the heat input to spacecraft debris and components in the free-molecular regime. Particles smaller than 1000 μm were assumed to come to thermal equilibrium instantaneously, while thermal response computer routines were used to determine the time-temperature histories of components. The significant particle parameters relating to the debris analyses are size, density, drag coefficient, accommodation coefficient, and surface emittance. The accommodation coefficient involves the amount of gas kinetic energy absorbed by the particle and depends on the body surface material as well as the gas molecular weight. The emittance determines the amount of energy radiated away. A limiting value of the ratio of accommodation coefficient to emittance of 0.39 was used, based on engineering judgment and a review of the literature [4]. Analyses were performed for three atmospheric models [5] and a range of entry angles. A spherical shape was used since supporting analyses indicated that it represented the most conservative case.

The spacecraft component free-molecular regime analyses involved additional thermal response calculations of the structure being considered. An example of thermal modeling that was required is that used for the antenna as shown in Fig. 2, where a simplified antenna geometry with the nodal subdivisions is pre-

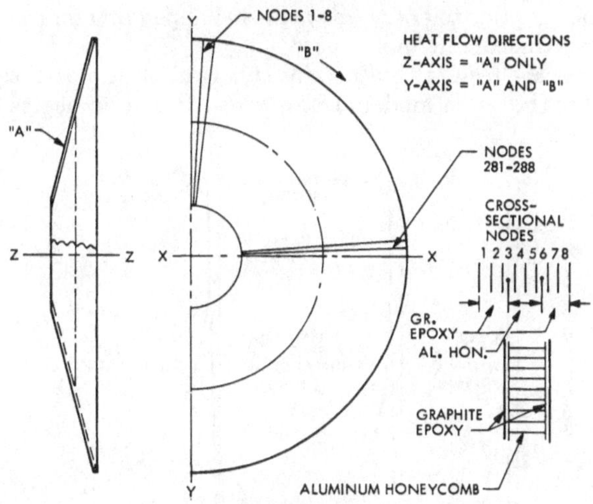

Fig. 2. Antenna geometry.

sented. The antenna dish structure is made up of the aluminum honeycomb core and two very thin graphite epoxy face sheets, one bonded to the front and the other to the back of the core by means of an epoxy resin. Since inadvertent entry may occur at any spacecraft attitude with respect to the velocity vector, the thermal response analyses were performed in two orientations: "face on" (along the Z axis) and "edge on" (along the Y axis).

It is assumed that the thermal responses of intermediate orientations are bounded by these two cases. The thermal model nodal subdivisions were based on the selected orientations. The thickness of the antenna was divided into eight layers conforming to the shape of the dish (see cross-sectional nodes sketch in Fig. 2). Each quadrant was divided into 36 sectors, thus forming 288 nodes per quadrant. For clarity, only the first and last two sectors in the (+Y, +X) quadrant are shown.

The analysis in the "face on" case required consideration of the heat flow in the "A" direction only because of concentric symmetry. In the "edge on" case, both heat flow modes, normal ("A") and tangential ("B"), were considered; thus all 288 nodes in the quadrant were used to obtain the actual temperature distribution. Only one-half of the dish was modeled because of circumferential symmetry. In the "edge on" case, part of the dish is exposed to the heat flux while the other part is in the shadow (no heating), and thus represents a heat sink. An appropriate number of nodes to account for this is used in the "edge on" analysis.

Similar analyses were performed for the thermal blankets which consisted of layers of aluminized Mylar. In addition, the temperature responses due to entry heating on the aluminum strut structure supporting the spacecraft appendages were determined.

2.3. Results

Ejecta and Disintegration Debris

The results of the particulate analyses indicate that ejecta released in the free-molecular regime, or prior to entry, may survive depending on particle properties, entry angle, and atmospheric model used.

Fig. 3 illustrates the results obtained for spherical plastic particles as a function of entry angle for the three model atmospheres. It also indicates that there is a

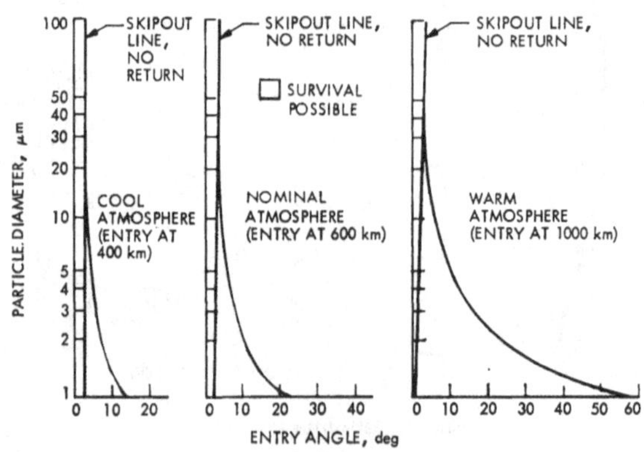

Fig. 3. Survival regions for spherical plastic particle containing micro-organisms.

minimum angle below which a particle will skip out of the atmosphere and not return. The area to the right of the shaded area represents the region where microbial "kill" would be achieved based on the lethality model [6] used in this analysis. In general, chances of survival increase as: (i) the particle size decreases, (ii) the entry becomes "shallow", and (iii) as the scale height of the assumed atmospheric models increases (i.e. warm atmosphere model).

Spacecraft Components

Analyses for the entires pacecraft, as it traverses the free-molecular regime and enters the early continuum, have shown that the thermal insulation blankets and the antenna dish disintegrate before entering the continuum flow, and that their disintegration should be completed in the early continuum. Also, the aluminum strut structure will melt away rapidly as soon as continuum flow is established.

Fig. 4 illustrates the identified disintegration stages of the outer planet spacecraft described in Fig. 1. The first stage, on the left, represents the spacecraft in outer space approaching the atmosphere of the planet. The spacecraft is still

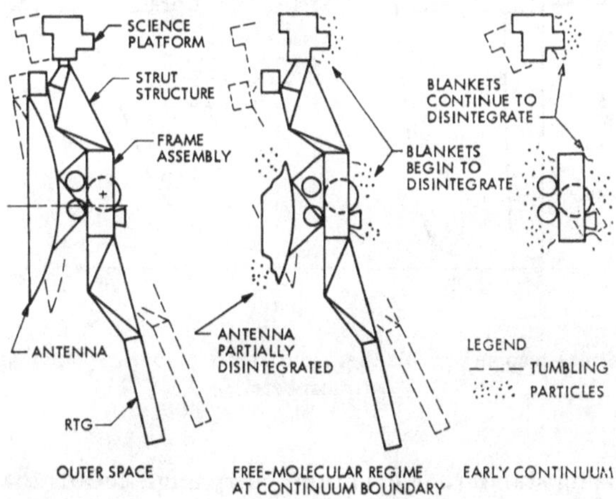

Fig. 4. Disintegration stages of outer planet spacecraft.

undisturbed but its flight may be stable or tumbling (dashed line indicates oscillation). The major parts of the spacecraft are identified in the figure. The second stage, in the middle, shows the spacecraft status when leaving free-molecular regime, i.e. at the continuum regime boundary. Here, the antenna is shown as partially disintegrated with the blankets also beginning to disintegrate. Small debris and particles, entrapped in these parts during manufacture, may be released at this time. These are indicated by a dotted shower. The third stage, on the right, portrays the situation in the early continuum. At this time, the aluminum strut structure has melted away, the antenna and other appendages have completely disintegrated, and the blankets are in the final stages of disintegration. In conclusion, what is left for the continuum flow entry analyses are the spacecraft science platform and the bus. The radioisotope thermoelectric generator has been eliminated from consideration, since its operational temperature is about 1000 °C in the center and over 350 °C on the outside; hence it was assumed to be inherently sterile. The disintegration described was caused primarily by thermal effects, because the maximum flow pressure generated on any part of the spacecraft did not exceed 0.01 atm; thus, the flow pressure could not contribute perceptibly to the spacecraft break-up.

Fig. 5 shows the typical results of thermal response of blankets during entry into the nominal Jupiter atmosphere; the abscissa is time from entry. The temperature histories for three angles are presented, and the limit temperature at which the blanket plastic evaporates is indicated. These data are representative of stage two of disintegration (Fig. 4). It is seen that the deeper the entry angle the shorter is the time of thermal response and the steeper is the temperature rise.

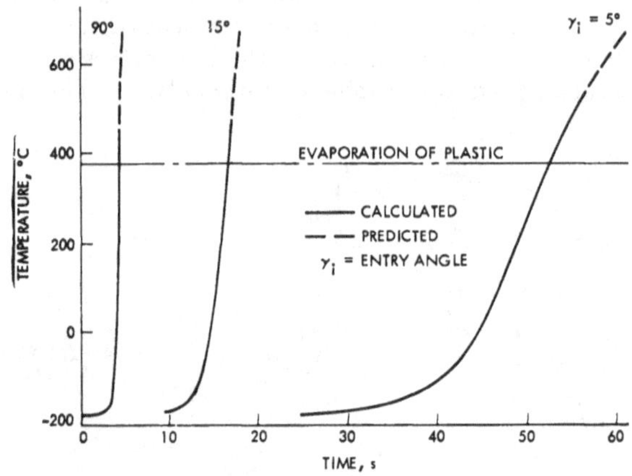

Fig. 5. Typical thermal response of blankets and struts of outer planet spacecraft, nominal atmosphere.

The 90° (−90° using standard convention) entry angle means that the spacecraft is heading straight in toward the center of the planet.

The results have shown that the antenna will only partially disintegrate before the continuum flow is reached, primarily due to relatively high heat resistance of the graphite epoxy face sheets. However, once disintegration has started, it will proceed rather rapidly in the continuum flow thermal environment.

3. Summary

This study has applied classical aerothermophysics techniques to determine the thermal responses of selected components of a typical Jupiter flyby spacecraft as well as ejecta and disintegration debris as they pass through the free-molecular flow regime and early continuum. Analyses for the entire spacecraft have shown that the thermal insulation blankets and the antenna are completely disintegrated by the early continuum. Large components, such as the spacecraft main support structure with the attached electronics and the science platform, will not receive sufficient heating in the free-molecular regime to cause major disintegration. However, complete disintegration is expected to occur in the continuum regime; analyses are being performed to determine the thermal response of these structures in the continuum. The analysis of spherical plastic particles indicates the existence of "survival corridors" which indicate that small diameter particles are the most likely to survive entry heating.

References

[1] W. Stavro and C. Gonzalez, Advanc. in Astronaut. Sci., The Outer Solar System, **29**, 465 (1971).
[2] B. L. Swenson, NASA Rep. TM X-2276, April 1971.
[3] B. L. Swenson, Body shape effects upon survivability during Jovian entry, Ames Research Center, Moffett Field, California, undated.
[4] R. C. Corlett, Astronautica Acta, **17**, 237 (1972).
[5] NASA Report SP-8069, 1971.
[6] A. R. Hoffman and J. A. Stern, in: Developments in Industrial Microbiology, **9**, 49 (1968), Society for Industrial Microbiology, Washington, D.C.

Life Sciences and Space Research XII — Akademie-Verlag, Berlin 1974

QUARANTINE CONSTRAINTS AS APPLIED TO SATELLITES

A. R. Hoffman, W. Stavro and C. C. Gonzalez

Jet Propulsion Laboratory, California Institute of Technology, Pasadena, Calif., USA

Plans for unmanned missions to planets beyond Mars in the 1970s include satellite encounters. Recently published observations of data for Titan, a satellite of Saturn, indicate that conditions may be hospitable for the growth of micro-organisms. Therefore, the problem of satisfying possible quarantine constraints for outer planet satellites was investigated. This involved determining the probability of impacting a satellite of Jupiter or Saturn by a spacecraft for a planned satellite encounter during an outer planet mission. Mathematical procedures were formulated which determine the areas in the aim-plane that would result in trajectories that impact the satellite and provide a technique for numerically integrating the navigation error function over the impact area to obtain impact probabilities. The results indicate which of the planned spacecraft trajectory correction maneuvers are most critical in terms of satellite quarantine violation.

1. Introduction

The United States is currently planning unmanned exploratory flyby missions to planets beyond Mars in the 1970s. One of the scientific goals of these missions is the investigation of the satellites of the outer planets. The trajectories of such missions therefore necessitate relatively close flyby encounter distances to the satellites. Concern for the biological preservation of these satellites has been voiced by the planetary quarantine community, in particular for such satellites as Titan of Saturn, where the probability of life is believed to be equivalent to, if not greater than, on the planet itself. As a result of this concern, a study was initiated to determine the implications of a satellite quarantine constraint on outer planet missions with close satellite encounters.

This paper presents the results of this investigation. The assumptions and method of analysis are presented first, and a parametric analysis is then performed for a typical Jupiter-Saturn mission with planned satellite encounters of Io at Jupiter and Titan at Saturn.

2. Method of Analysis

It will be assumed for the purposes of this paper that a quarantine constraint imposed on satellites of the outer planets will be violated if a spacecraft impacts a satellite. The likelihood of violation of the quarantine constraint from sources other than spacecraft impact (e.g. spacecraft debris) is assumed to be significantly smaller. Consequently, these other sources have been deleted from this investigation. The possibility of inadvertent impact of a satellite will always be present due to the inherent errors and uncertainties in the spacecraft navigation system. The theoretical *a priori* calculation of such accidental impact probabilities therefore becomes of primary importance in the analysis of the satellite quarantine problem for outer planet missions. The *a priori* probability of accidentally impacting a satellite can be determined if the following information is given: (i) a baseline spacecraft trajectory for a mission; (ii) the trajectory correction maneuver plan for such a mission; (iii) the quantitative errors that exist in these midcourse maneuvers as well as the injection maneuver; (iv) the orbit determination errors (e.g. planet and satellite ephemerides).

Given the information above, mathematical procedures now exist that can determine the probability of impacting the encounter planet. These tools, however, could not be utilized to determine the impact probabilities of satellites. The first step in this study, therefore, was to devise a procedure and develop the necessary tools to perform such a calculation. The procedure used in this analysis is the following:

(i) Determine the areas in the aim-plane of the encounter planet that would result in trajectories which impact the satellites of the planet.

(ii) Integrate the probability distribution resulting from the navigation errors over these areas to determine the probability of satellite impact.

A theoretical formulation was devised and a computer program developed to perform the first step in the procedure outlined above. The results showed that these areas were approximately elliptical in most practicable cases. Because of this result, the available conventional tools for performing step (ii) in the procedure above could not be used, since they integrated the probability distribution over a *circular* area (which represents the impact area for a *planet*). Another new formulation was therefore developed and programmed to perform the integration over an elliptical area.

With these software tools developed, it becomes possible to determine the satellite impact probability and thus the satellite quarantine implications for satellite encounter missions.

3. Application to a Sample Mission

3.1. Mission Characteristics

The mission selected for analysis is a typical Jupiter-Saturn mission. The particular trajectory analyzed is one which encounters the satellite Io at Jupiter and the satellite Titan at Saturn. The trajectory correction maneuver plan for such a mission is depicted in Fig. 1. Injection plus six trajectory correction maneuvers are planned, three during the earth-Jupiter phase and three during the Jupiter-Saturn phase. The times and purposes for these maneuvers are given in Fig. 1.

For such missions, an important required parameter is the 1σ total delivery error in the navigation system. Representative values are given in Table 1 and will be used here as sample numbers in order to investigate their implications on satellite quarantine. These errors are given in the Jupiter aim-plane for the injection plus the first three trajectory correction maneuvers, and in the Saturn aim-plane for the Jupiter-Saturn leg maneuvers as well as the last pre-Jupiter maneuver (see Fig. 1). As shown in the results, both the size and the ellipticity of these values are important.

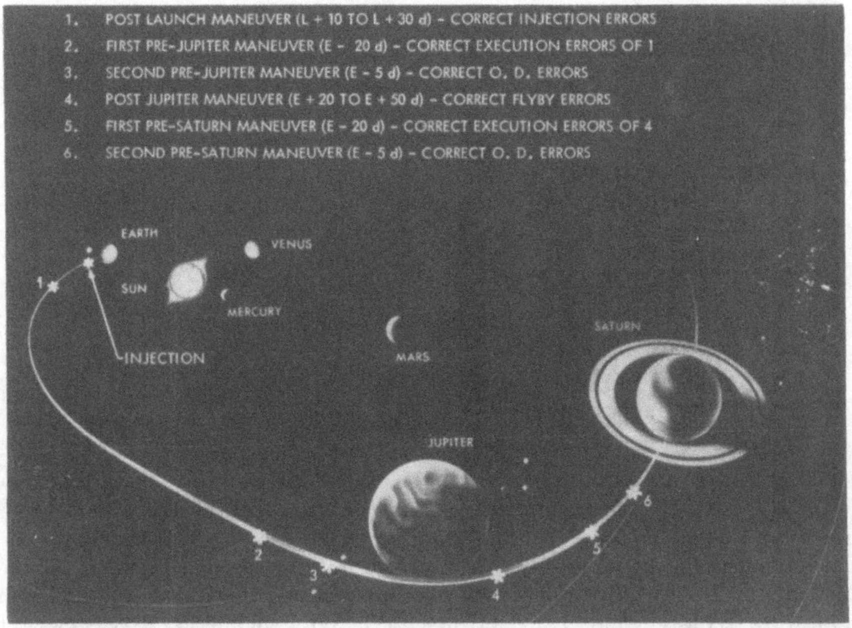

Fig. 1. Trajectory correction maneuver plan for Jupiter-Saturn mission (O.D. is orbit determination).

Table 1

Total Navigation Delivery Errors (1σ)

Maneuver	Jupiter aim-plane			Saturn aim-plane		
	$\sigma(\vec{B}\cdot\vec{R})$ (km)	$\sigma(\vec{B}\cdot\vec{T})$ (km)	Correlation coefficient	$\sigma(\vec{B}\cdot\vec{R})$ (km)	$\sigma(\vec{B}\cdot\vec{T})$ (km)	Correlation coefficient
Injection	120000	521000	0.176			
Maneuver 1	2302	11213	0			
Maneuver 2	1200	823	0.08			
Maneuver 3	1100	436	0.08	402000	201000	−0.985
Maneuver 4				4037	4233	−0.137
Maneuver 5				651	2123	−0.043
Maneuver 6				414	2000	−0.083

3.2. Application Procedure

Using the developed tools and the trajectory and navigation characteristics, the probability of satellite impact was calculated. The probability of satellite impact was found to be a function of two variables: the closest approach distance r and orientation angle α. These variables are depicted in the planet aim-plane geometry shown in Fig. 2. The orientation angle represents the location of the

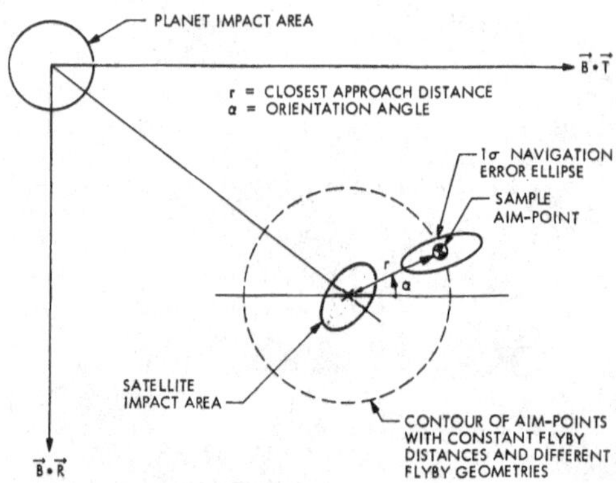

Fig. 2. Planet aim-plane (not to scale).

trajectory aim-point with respect to the center of the satellite in ecliptic space. Although the closest approach distance r of the spacecraft to the satellite for Io is fixed relative to the planet by the geometry of the nominal trajectory, it can be allowed to change considerably with respect to the satellite without severely handicapping the mission. For Titan, the distance is determined by science and engineering requirements, since no specified flyby geometry is necessary for gravitational assist, as it is for Jupiter.

In the analysis, various values of r were used and, for each value, α was varied from 0 to 360°. In doing so, the probability of impact changes as α is changed, even though r is fixed. An important observation is that the magnitude of the changes is larger when either the satellite impact area or the navigation errors become more elliptic. For the special case where the satellite impact area and the 1σ navigation errors are circular, the impact probability becomes a function of r only and not of α.

Since impact probability changes as α varies (for a fixed r), one is interested from a quarantine point of view in knowing the maximum value. These values are listed in Table 2 both for Io at Jupiter and for Titan at Saturn.

3.3. Interpretation of Results

The resulting impact probabilities have some very interesting and important implications for satellite quarantine. To determine whether satellite quarantine was violated, it was necessary to assume a satellite quarantine constraint (1×10^{-5}) and determine whether the contamination probability was equal to or greater than the constraint. In interpreting the impact probabilities in terms of the contamina-

Table 2
Maximum Probabilities of Satellite Impact

Maneuver	Closest approach (km)	Probability of satellite impact (maximum) Io at Jupiter	Titan at Saturn
Injection	Any	$<10^{-6}$	
Maneuver 1	5000	5×10^{-2}	
	10000	3.7×10^{-2}	
	15000	2.3×10^{-2}	
	20000	1.1×10^{-2}	
	30000	1.6×10^{-3}	
	40000	1×10^{-4}	
Maneuver 2	5000	2.4×10^{-3}	
	$\geqq$ 8000	$<10^{-6}$	
Maneuver 3	5000	1.3×10^{-3}	$<10^{-6}$
	$\geqq$ 7000	$<10^{-6}$	$\ll 10^{-6}$
Maneuver 4	8000		4×10^{-2}
	10000		1.8×10^{-2}
	12000		6.1×10^{-3}
	14000		1.9×10^{-3}
Maneuver 5	8000		5×10^{-3}
	10000		2×10^{-4}
	$\geqq$12000		$<10^{-6}$
Maneuver 6	8000		2.8×10^{-3}
	10000		1×10^{-4}
	$\geqq$12000		$<10^{-6}$

tion probabilities (i.e. of violating a quarantine constraint), it must be noted that the stated values do not take into account the probability of being able to perform a corrective maneuver. Typically, this would decrease by approximately two orders of magnitude [1] the values listed in Table 2 for the injection maneuver and trajectory correction maneuvers 1 through 5. Maneuver 6, since it is the last maneuver in the mission, cannot be adjusted by the probability of performing a corrective maneuver because none is planned.

The determination of the contamination probabilities gives the following important results:

(i) The injection maneuver, which can be a critical maneuver relative to violating the Jupiter quarantine [1], does not violate a typical satellite quarantine on a Jupiter satellite.

(ii) Maneuver 3, which was critical in terms of violating a Saturn quarantine [1], does not violate a Saturn satellite quarantine.

(iii) Maneuvers 2 and 3 for Jupiter and maneuver 5 for Saturn (see Fig. 1) would violate a sample satellite quarantine constraint of 10^{-5} only for very close satellite encounter distances (less than 5000 km) and for specific flyby geometries.

(iv) Maneuver 1 would violate an Io quarantine of 10^{-5} for flyby distances of up to 40000 km; however, the impact probability can be reduced to negligible values by changing the satellite flyby geometry.

(v) Maneuver 4 seems to be the most critical since it would violate a Titan quarantine constraint of 10^{-5} for flyby distance of up to 11000 km regardless of flyby

geometry. The reason for this is the circularity of both the satellite impact area and the navigation error ellipses.
(vi) Maneuver 6 would violate a Titan constraint of 10^{-5} for Titan flyby distances of less than 12000 km for specific flyby geometries. The reason for this is that no subsequent maneuvers can be relied upon to correct an impact trajectory.

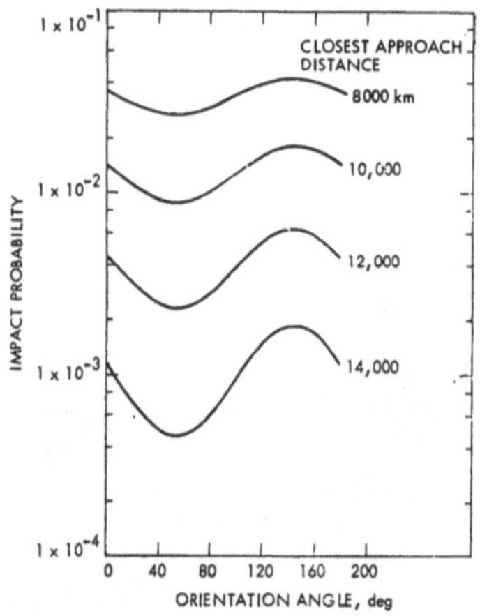

Fig. 3. Titan impact probabilities for various flyby designs (maneuver 4).

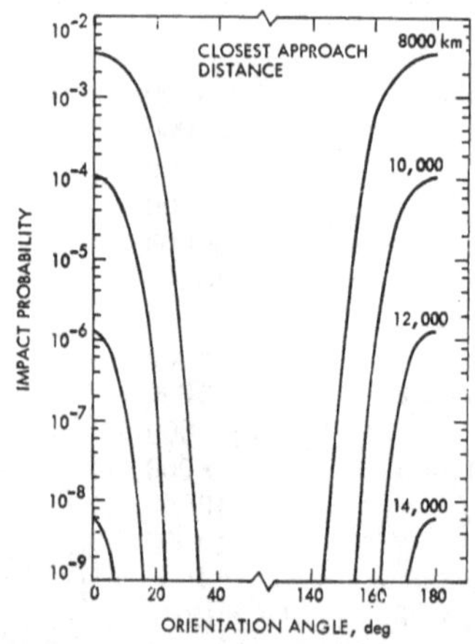

Fig. 4. Titan impact probabilities for various flyby designs (maneuver 6).

As mentioned before, the probability of impact values listed in Table 2 is strongly dependent on the eccentricity of the navigation error ellipse. To illustrate this dependence, Fig. 3 shows the variation of Titan impact probability versus orientation angle for maneuver 4 (fairly circular error ellipse), and Fig. 4 shows the variation for maneuver 6 (highly elliptic error ellipse).

4. Summary and Conclusions

The implications of a satellite quarantine constraint on a typical outer planet satellite encounter mission were studied from the point of view of satellite impact probabilities. Two important conclusions should be drawn from this effort: (i) significant differences exist between planetary and satellite quarantine implications on outer planet missions. For example, trajectory correction maneuvers that result in a possible satellite quarantine violation are different from those violating *planetary* quarantine. (ii) Analytical tools for determining the probability of a spacecraft impacting a satellite are now available for application to any set of trajectory and navigation characteristics for satellites.

References

[1] W. Stavro and C. Gonzalez, Advanc. in Astronaut. Sci., The Outer Solar System, **29**, 465 (1971), American Astronautical Society, Tarzana, Calif.

INDEX OF AUTHORS